Manfred Tschabitscher
Clemens Klug

Endoscopic Anatomy of the Middle Ear

Springer-Verlag Wien GmbH

Univ.-Prof.Dr. Manfred Tschabitscher
Dr. Clemens Klug
Department of Anatomy, University of Vienna,
Study Group Microsurgical & Endoscopic Anatomy,
Vienna, Austria

Originally published by Springer-Verlag Wien New York in 2000

Typesetting: Stefan Kindel, D-67487 Maikammer

Printed on acid-free and chlorine-free bleached paper

SPIN: 10574970

With 246 coloured Figures

CIP data applied for

ISBN 978-3-211-82973-8 ISBN 978-3-7091-6319-1 (eBook)
DOI 10.1007/978-3-7091-6319-1

Endoscopic anatomy is not conventional anatomy "re-invented". It simply describes known anatomic details looked at under different viewing conditions: Through the scope structures are viewed at an angle other than that of the operating microscope. Consequently, the segments in the field of view are different as are the relative sizes of the structures seen. Thanks to the superior lighting, details can be better discriminated. In addition, the scope provides a characteristic "keyhole" view. It only sees structures directly ahead of it or rather ahead of its front lens, while structures it has already passed (e.g. those along the approach chosen) lie outside its field of view and are no longer seen, particularly with flexible scopes.

Two general factors are essential for endoscopy, (1) light and (2) a space in which the light can spread. In the ear this space is naturally provided by the tympanic cavity.

As whatever is done in medicine should produce an optimal therapeutic effect with minimal iatrogenic trauma [PERNECZKY, 1993], the use of the endoscope in the middle ear "space" is an inevitable and logical choice.

The idea of gaining access to the tympanic cavity surgically through the mastoid process was born in the middle of the 17th century [RIOLAN]. Since the days of SCHWARTZE [1873] antrotomy and mastoidectomy have been standard procedures in ear surgery. Paracentesis for entering the middle ear was first performed by BONNAFONT [1934]. MER [1967] introduced middle ear endoscopy through a transmeatal approach and KIMURA [1989] described a transtubal approach for exploring the middle ear structures.

As scopes are perfected technically (light intensity, heat generation, flexibility, steerability) and as diagnostic accuracy and patient selection are improved by high-resolution imaging techniques (CT, MRI), endoscopy will, no doubt, become more widely used in otology.

Today scopes "still" mainly serve diagnostic purposes. Their therapeutic uses are confined to endoscopically assisted microsurgery, while true endoscopic surgery has not yet become a reality. But this will dramatically change in the future, as existing instruments are improved and new technologies are developed, e.g. lasers (for cutting and drilling), water jet dissection (for transecting tissues), ultrasound (for real-time navigation), internal-pressure-controlled micro-instruments (so-called GIORDANO instruments), etc.

The concept of "minimal invasiveness" has been dealt with at considerable length in the section on **Transtubal Endoscopy**, because we are convinced that,

if indications are defined by stringent criteria, the technique will gain an important place particularly in pediatric otology. Its clinical indications can naturally not be covered in this volume.
As the working space available is limited and the structures in it are very delicate, the iatrogenic trauma associated with endoscopy must be minimized at all cost. This requires a profound understanding of anatomy. This volume will help the readers acquire or broaden it.

April 2000 *M. Tschabitscher* *C. Klug*

Acknowledgements

The long tradition of the Institute of Anatomy at the University of Vienna to collaborate with clinical services, which was begun by HYRTL and continued by ZUCKERKANDL, TANDLER, PERNKOPF, V. HAYEK, GISEL, KRAUSE and FERNER, generated an atmosphere which still is singularly conducive to research in gross anatomy. We gratefully acknowledge the generous support of our work by the Medical School of the University of Vienna. But we are also indebted to AESCULAP, STORZ and ZEPPELIN companies for providing us with instruments. Our special thanks go to TONI NOBLES from NOBLES-LAI INC., who made the Hawk Vision System with its ultrathin flexible scopes available to us. Without it endoscopic work of this quality would not have been possible.

With gratitude we recognize the assistance of Balas Fabinyi, M.D., senior resident at the ENT Service, Krems Hospital, in clinical matters.

We are very grateful to Mag. Krista Schmidt for preparing the English-language manuscripts.

Last, but not least, we owe a vote of thanks to all those who, following a long tradition in Vienna, donated their bodies to the Institute of Anatomy for teaching and research. Without them scientific work in anatomy would be unthinkable.

The endoscopes used for these studies are not equipped with a working channel and an irrigation/suction tube so that their outer diameter is no larger than needed for the light cable and the lens.

I. Endoscopes

Rigid scopes:

a) 0°, 1.9 mm outer diameter, 60 mm length (Aesculap, Tuttlingen, Germany)
   0°, 4 mm outer diameter, 150 mm length, angled attachment of the light cable and eyepiece (Aesculap, Tuttlingen, Germany)

c) 30°, 1.9 mm outer diameter, 100 mm length (Karl Storz, Tuttlingen, Germany)

d) 30°, 2.7 mm outer diameter, 100 mm length (Karl Storz, Tuttlingen, Germany)

e) 30°, 4 mm outer diameter, 150 mm length, angled attachment of the light cable and eyepiece (Aesculap, Tuttlingen, Germany)

f) 60°, 1.9 mm outer diameter, 180 mm length (Richard Wolf, Kittlingen, Germany
   70°, 4 mm outer diameter, 160 mm length (Karl Storz, Tuttlingen, Germany)

Flexible non-steerable scope:
0.94 mm outer diameter, 1070 mm length (Hawk Vision, Nobles-Lai Inc., California, USA)

Flexible steerable scope:
0.8 mm outer diameter, 600 mm length (Micromed, Herne, Germany)

II. Light Fountain:

Hawk Vision System: Hawk Vision, Nobles - Lai Inc., California, USA
Cold Light Fountain Typ CLV - S: Sony, Austria
Flash generator - 600: Karl Storz, Tuttlingen, Germany

III. Video Recording System:

Umatic - High band : Sony, Austria
Betacam : Sony, Austria
S - VHS : Sony, Austria

IV. Camera Equipment:

Contax 167 MT: Yashika Co.Ltd., Vienna, Austria
Olympus OM 4 TI with an Olympus SM-ER 2 adapter: Olympus, Austria
Ricoh XR-X: Ricoh, Austria

V. Photographic Film:

Kodak Ektachrome 400X Professional
Kodak Ektachrome 100 Professional

VI. Operating Microscope:

OPMI-6 cfc: Karl Zeiss GmbH, Vienna, Austria

A Transmeatal approach

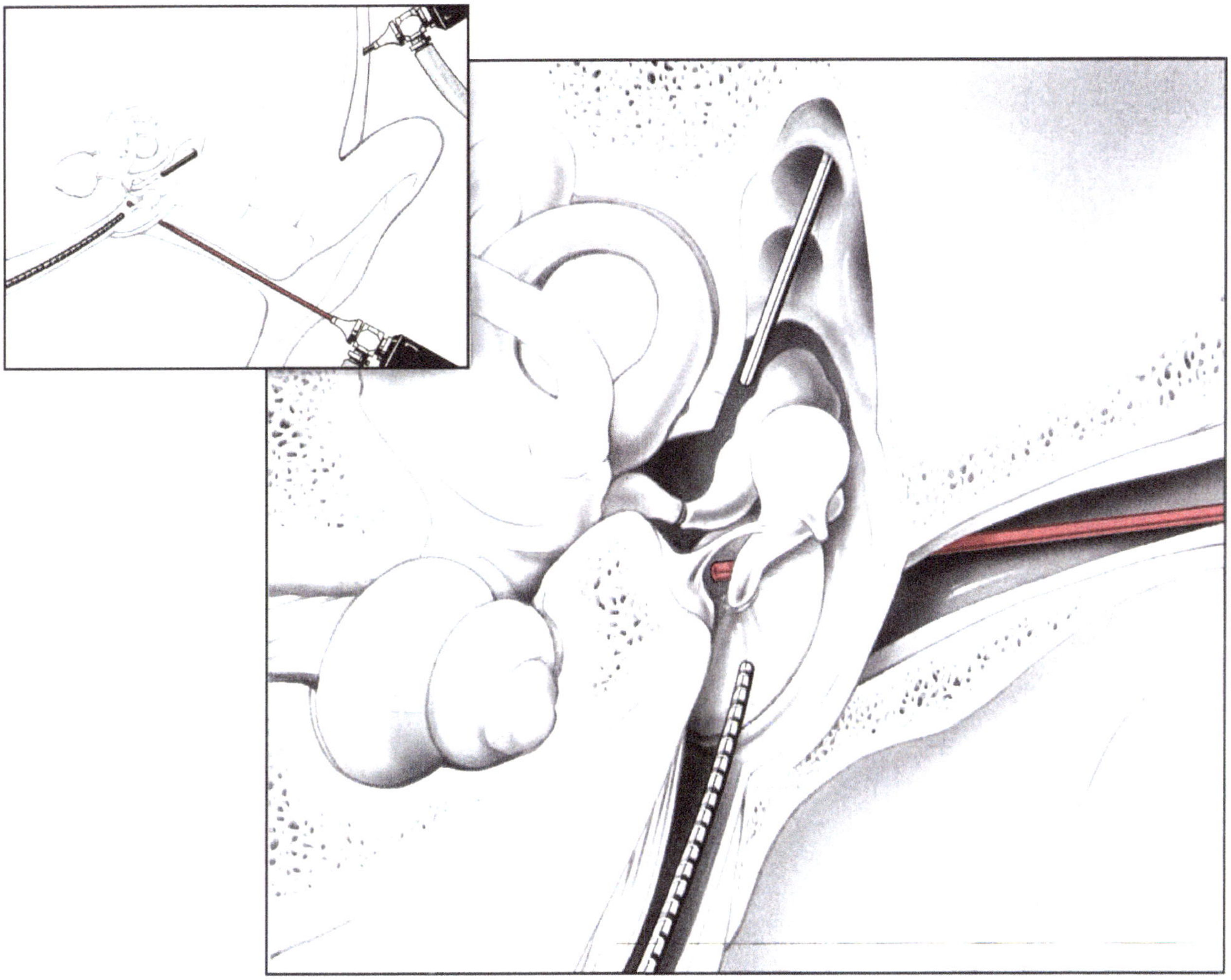

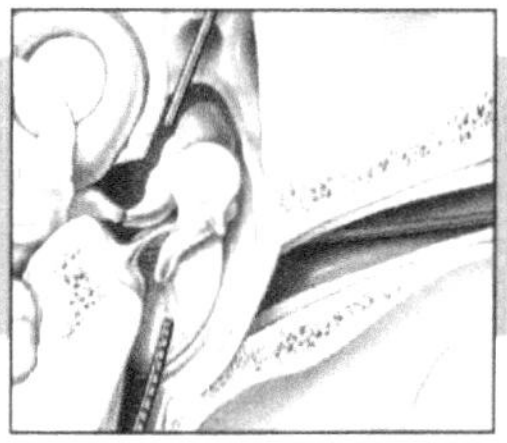

Transmeatal approach to the middle ear

Although the tympanomeatal flap technique is much more invasive than the transtympanic approach, the large access it provides without major trauma will accept relatively large scopes (up to an outer diameter of 4 mm) so that an excellent overview is obtained. If smaller scopes (1.9-2.7 mm) are used, another instrument can be introduced at the same time, provided the external auditory canal is not excessively kinked and malformations (e.g. atresia, strictures, hyperostosis) are absent.

Using a sickle knife, incisions are made in the posterior wall at 1 and 6 o'clock from the tympanic membrane towards lateral. These incisions are connected by another cut in the posterior wall of the auditory canal, which is made with a round cutting knife. The resultant flap of the lining of the auditory canal is elevated from the bone with an elevator to expose the tympanic ring, which is luxated out of the tympanic sulcus. Together with the elevated lining of the auditory canal the posterior half of the tympanic membrane forms a tympanomeatal flap which can be reflected anteriorly. To close the defect the flap is simply brought back into its original position.

Depending on the angle of the rigid scope, virtually any region of the middle ear is seen, except for the inner aspect of the lateral wall ("keyhole" effect!). The use of a transmeatal scope for monitoring the progress of a transtubal scope is discussed at length in Section IV.

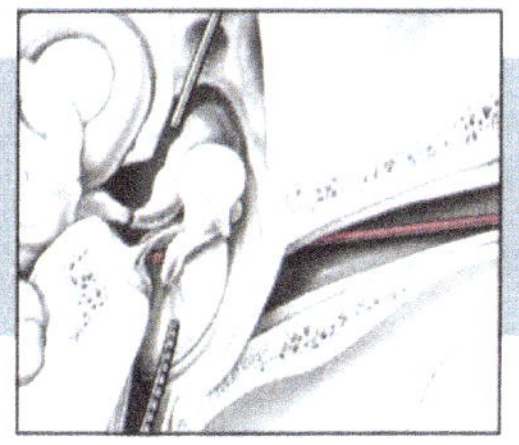

A Transmeatal approach

4 mm rigid scope

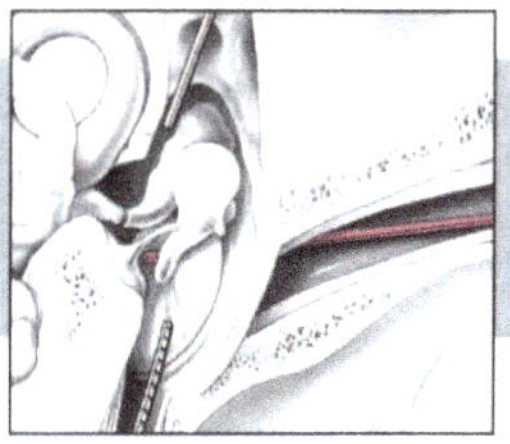

A 1.1 Approach to the tympanic cavity with a tympanomeatal flap technique

Fig. 1 Right tympanic membrane (rigid scope, 4 mm, 0°).

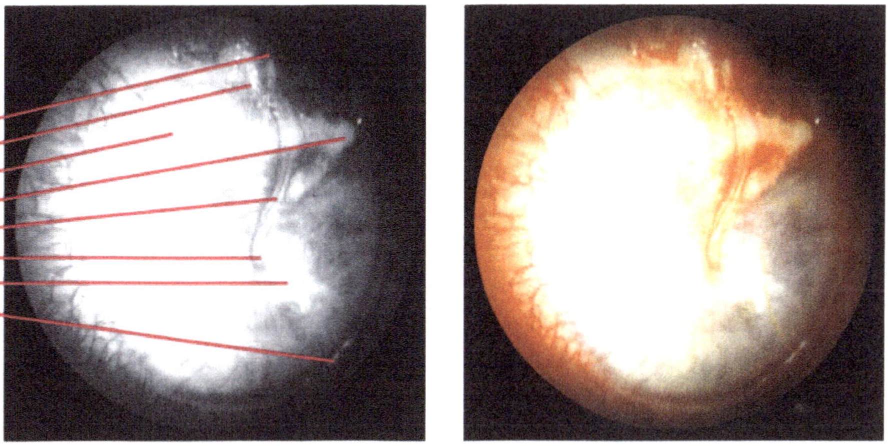

Fig. 2 View into a right tympanic cavity after reflecting the tympanomeatal flap anteriorly (rigid scope, 4 mm, 0°).

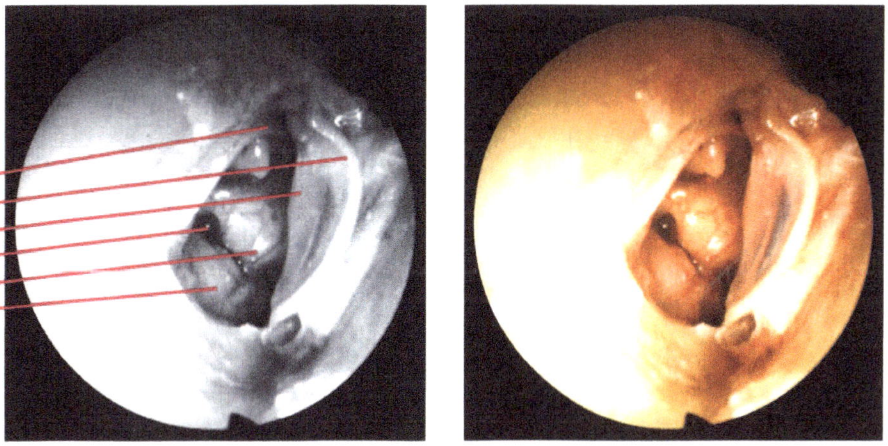

Fig. 3 Scope further advanced: The central structures of the middle ear are seen opposite the tympanic membrane (rigid scope, 4 mm, 0°).

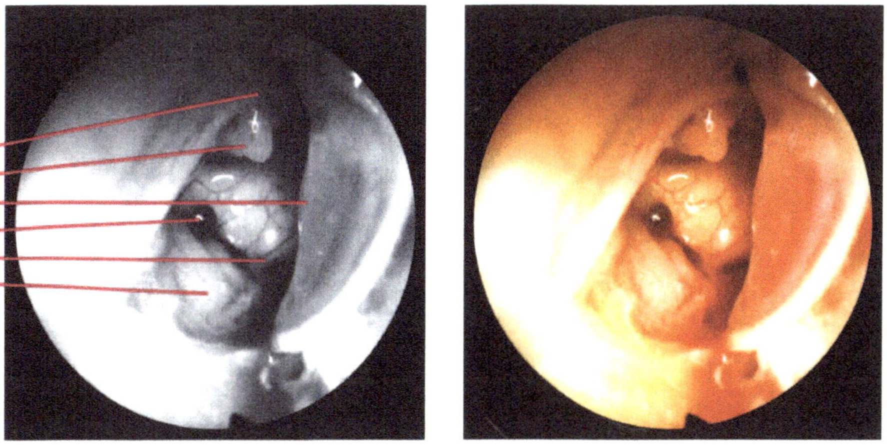

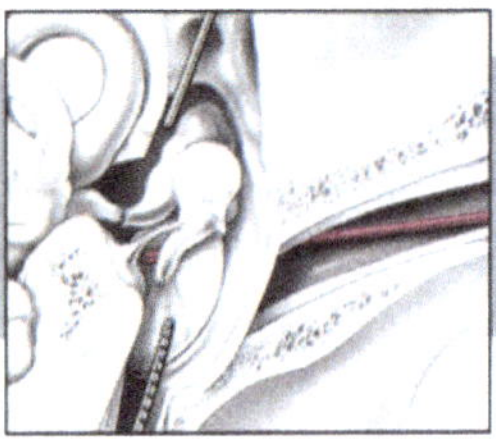

A 1.2 Following the course of the chorda tympani

Moving the scope from anterior-superior to posterior-superior

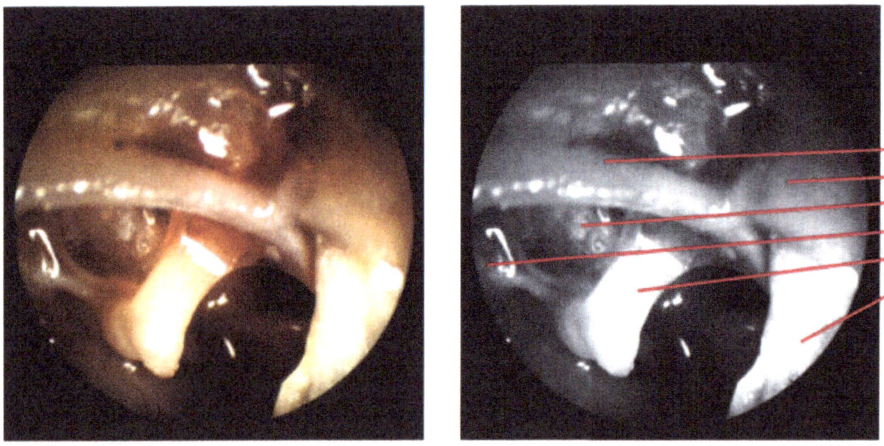

Fig. 4 Looking anteriorly and superiorly the right chorda tympani between incus and malleus is seen (rigid scope, 4 mm, 70°, right ear).

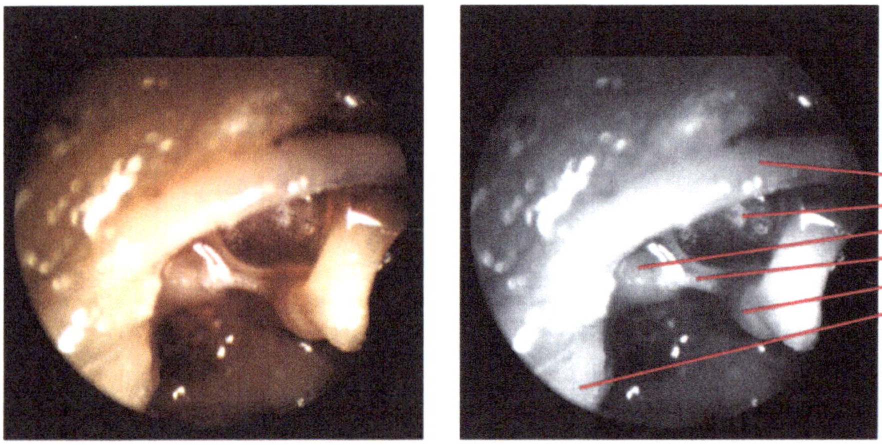

Fig. 5 Looking posteriorly and superiorly the chorda tympani is seen, as it leaves the chordal eminence (rigid scope, 4 mm, 70°).

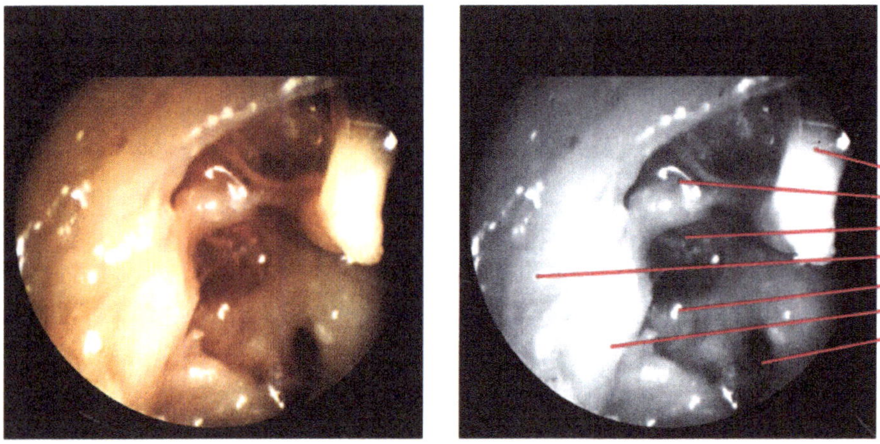

Fig. 6 View focused on chordal eminence (rigid scope, 4 mm, 70°).

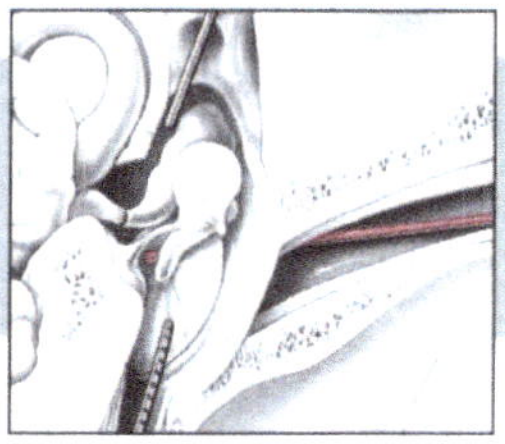

A 1.3 View into the anterior epitympanic recess

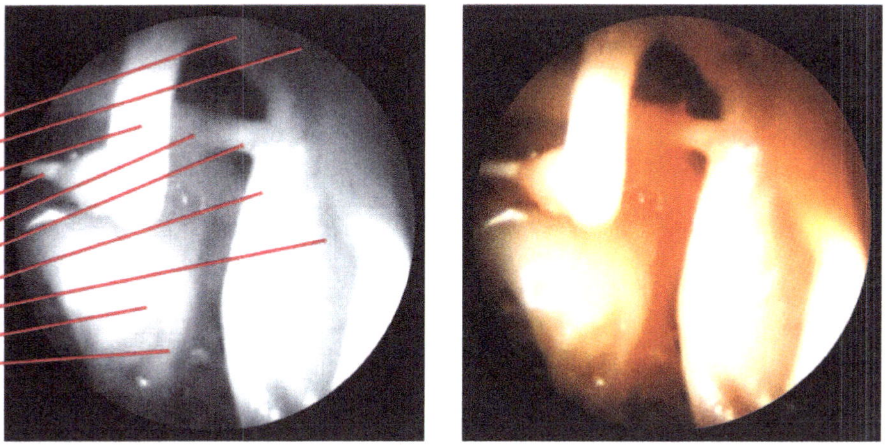

Fig. 7 View through gap between incus and malleus (rigid scope, 4 mm, 30°, right ear).

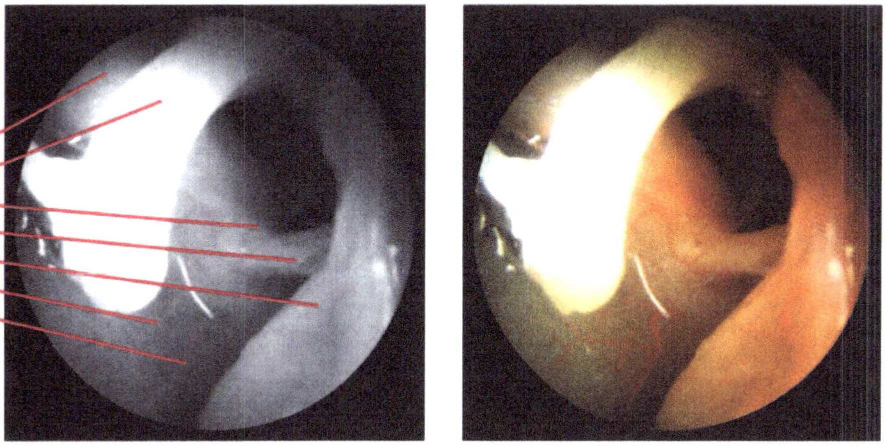

Fig. 8 The scope is moved on toward the cochleariform process (rigid scope, 4 mm, 30°).

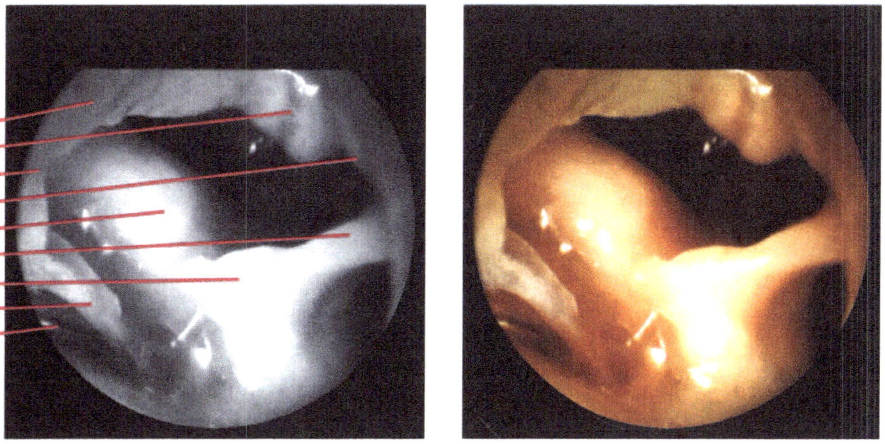

Fig. 9 View looking upward into epitympanic recess (attic) with incudomallear joint seen from below (rigid scope, 4 mm, 70°).

A 1.4 Overview of the protympanum by coaxial rotation from the top down

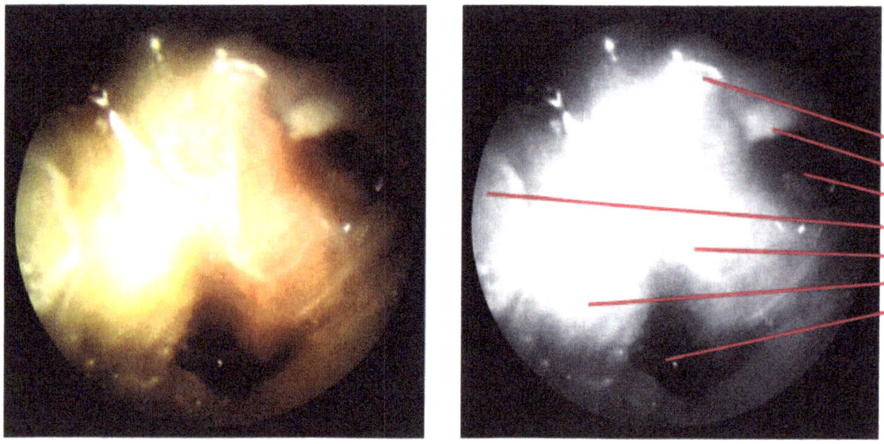

Fig. 10 Looking forward the region of the protympanum is seen (rigid scope, 4 mm, 30°, right ear).

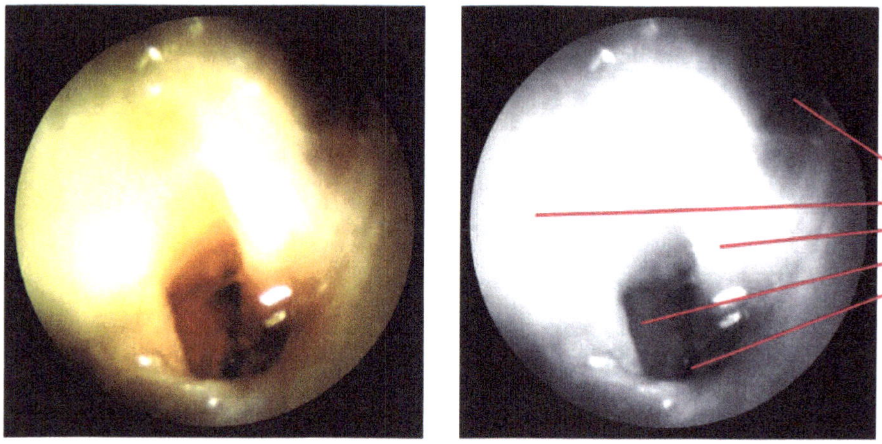

Fig. 11 Boundaries of the tympanic opening of the EUSTACHIAN tube (rigid scope, 4 mm, 30°, looking forward).

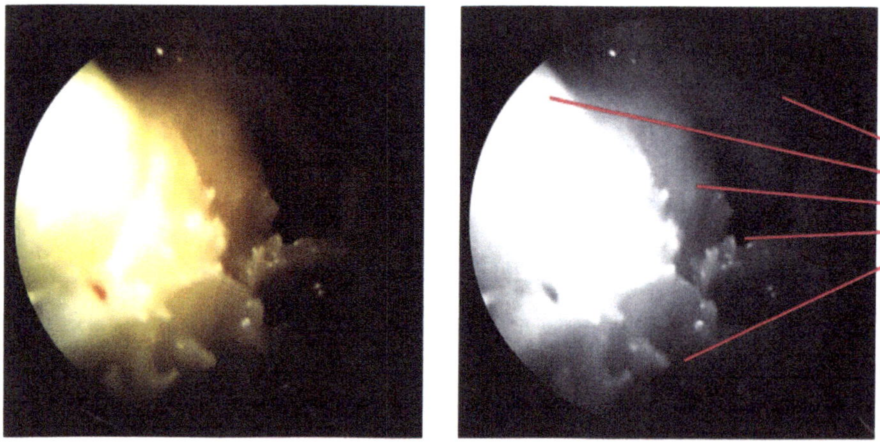

Fig. 12 Floor of the protympanum (rigid scope, 4 mm, 30°, looking forward down).

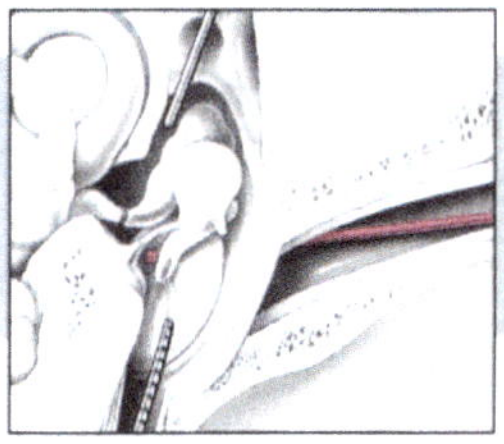

A 1.5 The central mesotympanum seen through 70°, 30°, 0° scopes – comparative views

(with nearly the same distance to promontory)

Fig. 13 Right mesotympanum in the center, incus and facial canal are also seen (rigid scope, 4 mm, 70°, right ear).

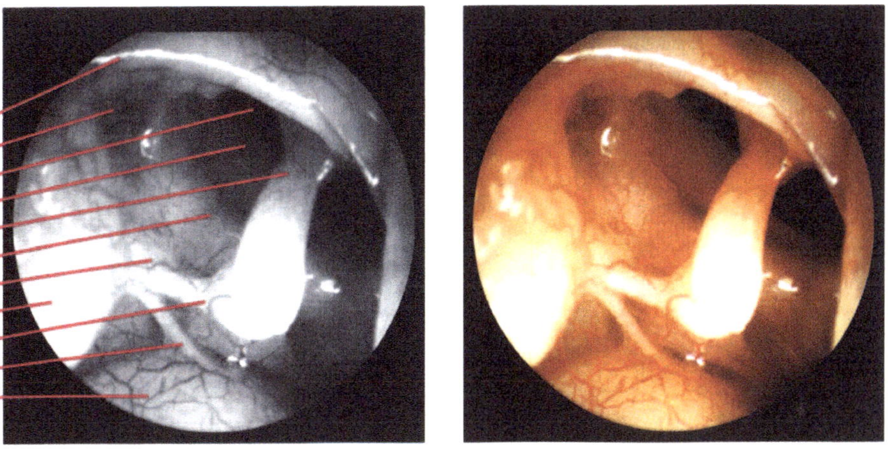

Fig. 14 Focusing on promontory, round window niche and incudostapedial joint (rigid scope, 4 mm, 30°).

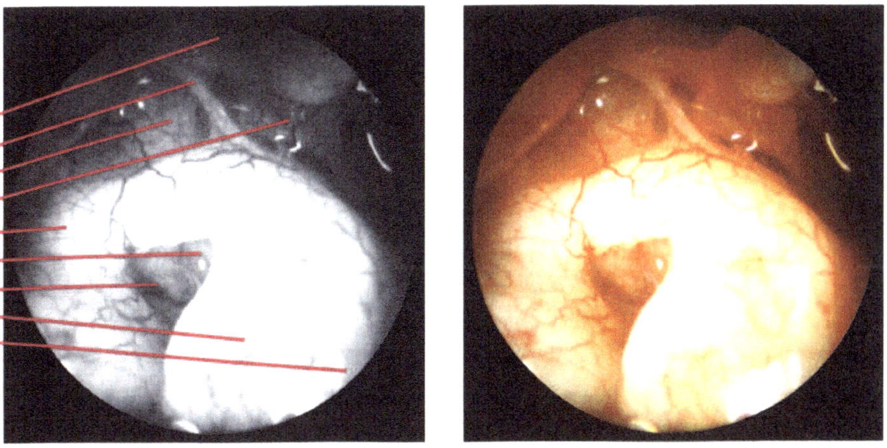

Fig. 15 Looking at the promontory and round window nice (rigid scope, 4 mm, 0°).

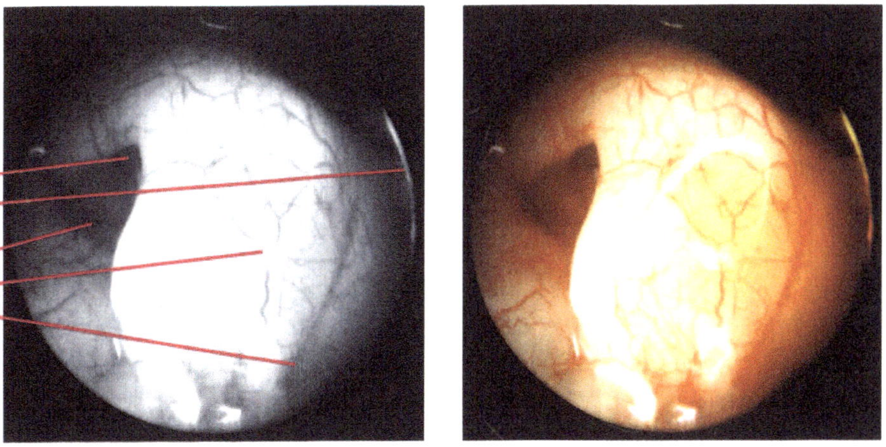

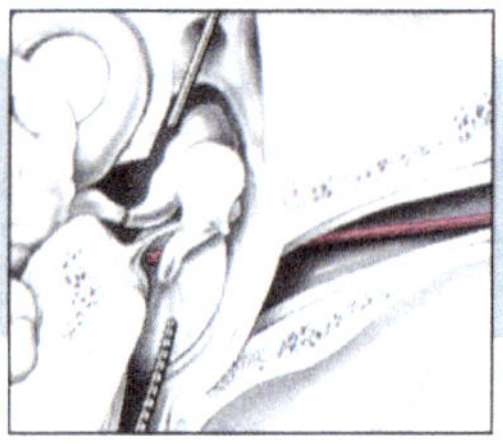

A 1.6 Incus

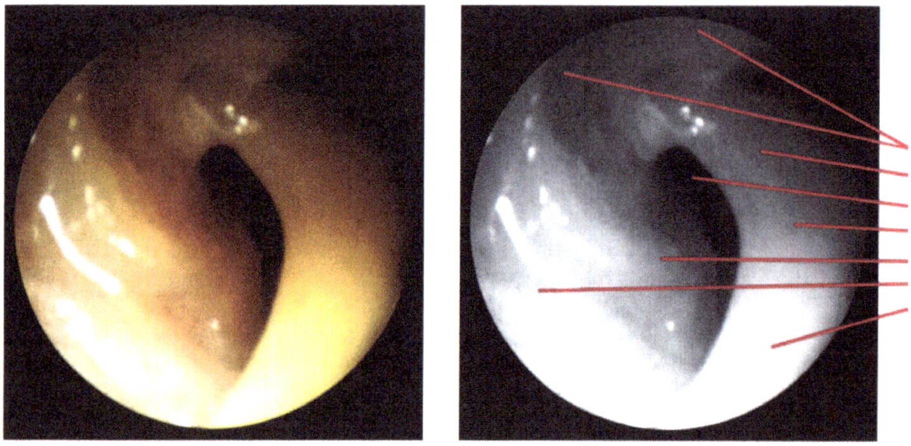

Fig. 16 Looking posteriorly and up toward short limb of incus with ligament (rigid scope, 4 mm, 70°, right ear).

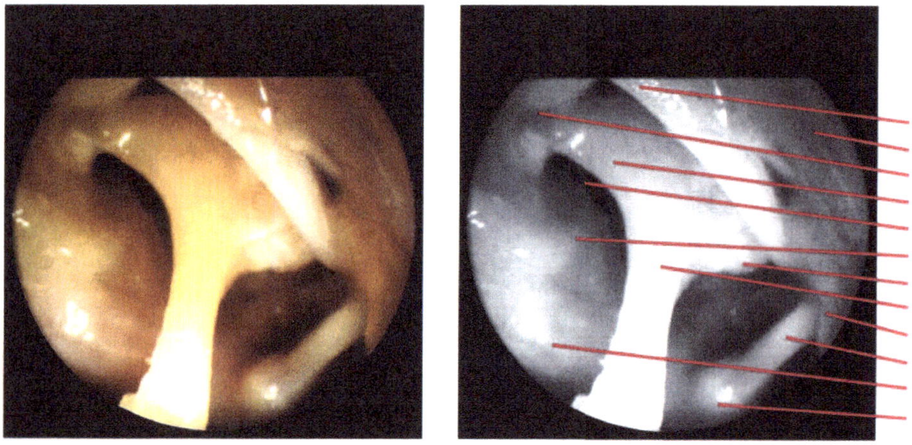

Fig. 17 Body of incus and chorda tympani (rigid scope, 4 mm, 70°, looking up).

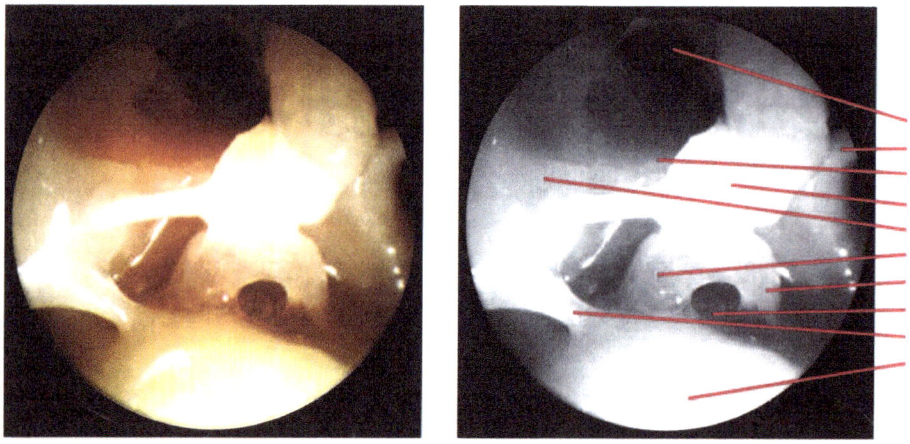

Fig. 18 View of incudostapedial joint (rigid scope, 4 mm, 30°, looking up).

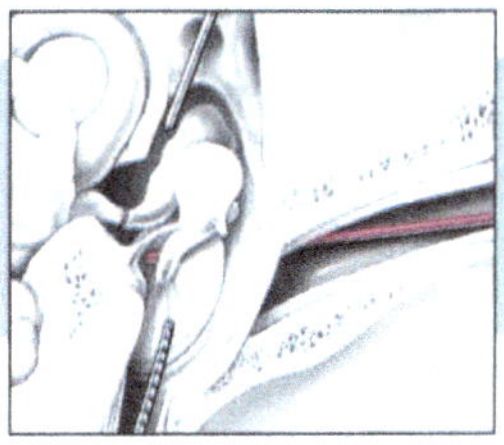

A 1.7 Stapes and tendon of stapedius muscle

Fig. 19 Looking up and focusing on the stapes suprastructure (rigid scope, 4 mm, 70°, right ear).

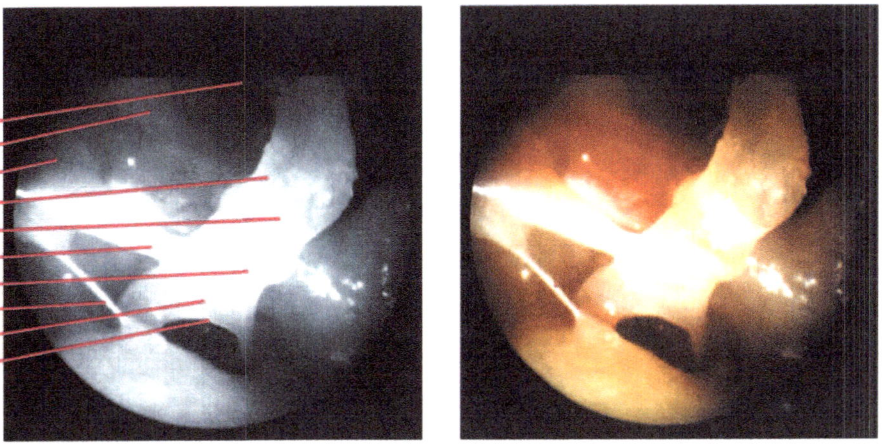

Fig. 20 Looking posteriorly and up (rigid scope, 4 mm, 70°).

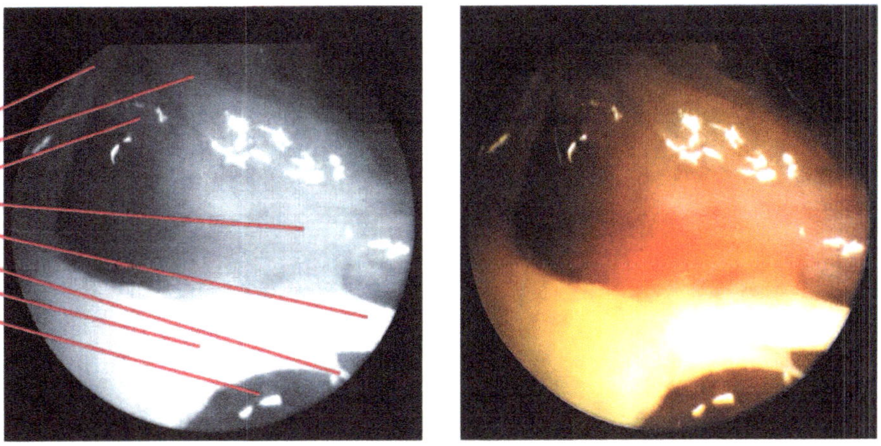

Fig. 21 Centering on pyramidal eminence gives a complete overview of all of the posterior "middle ear recess" (rigid scope, 4 mm, 70°).

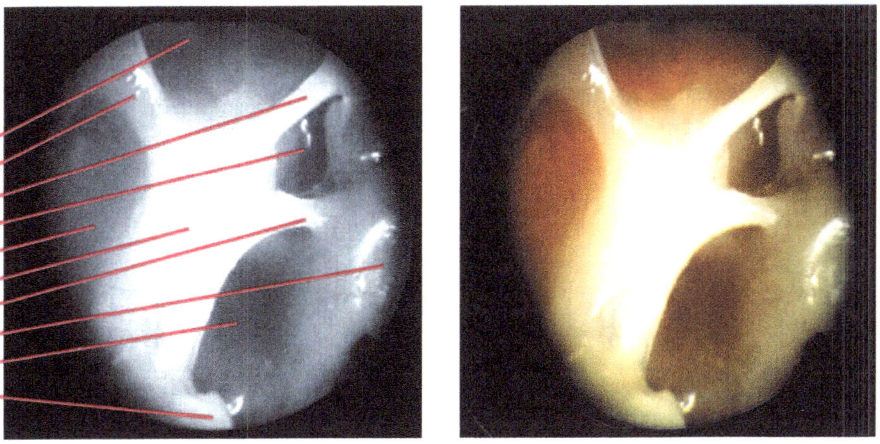

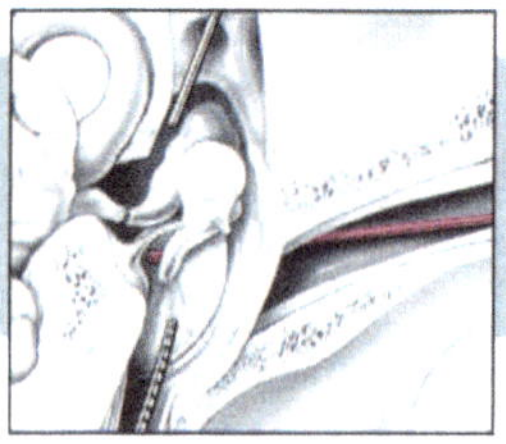

A 1.8 Mucosal fold surrounding the ossicular chain

Seen through a 70° scope on a straight way to the promontory

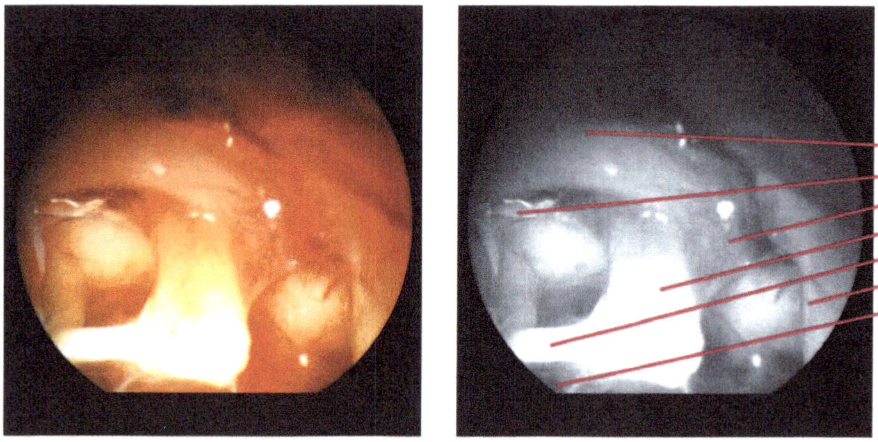

Fig. 22 The tip of the scope is in the plane of the tympanic membrane looking up (rigid scope, 4 mm, 70°, right ear).

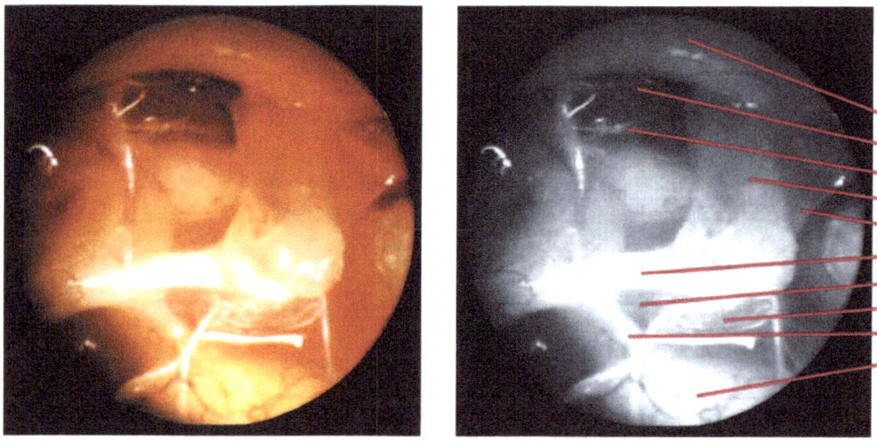

Fig. 23 The scope is advanced 2 mm beyond the tympanic membrane (rigid scope, 4 mm, 70°).

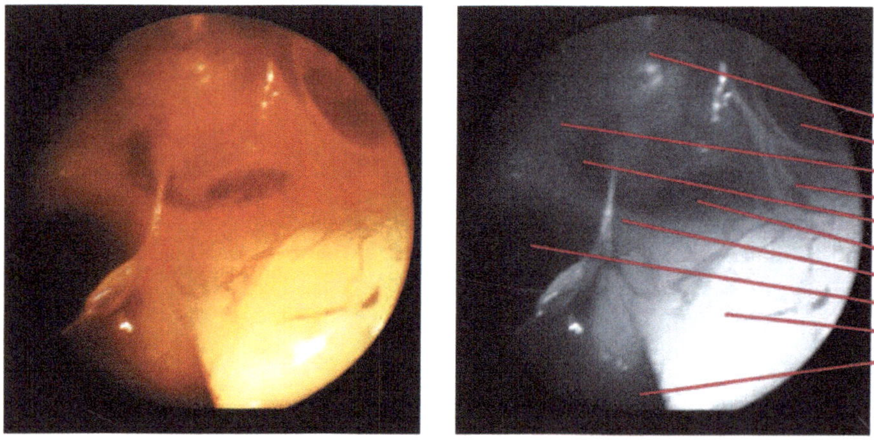

Fig. 24 The tip of the scope is close to the promontory (rigid scope, 4 mm, 70°).

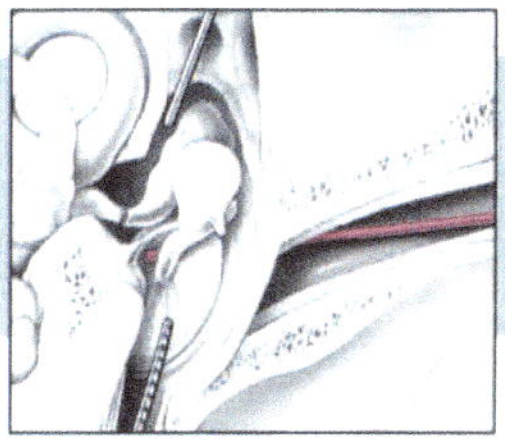

A 1.9 Posterior wall of tympanic cavity

Seen by rotating the scope from superiorly to inferiorly

Fig. 25 Superior portion of the posterior wall (rigid scope, 4 mm, 70°).

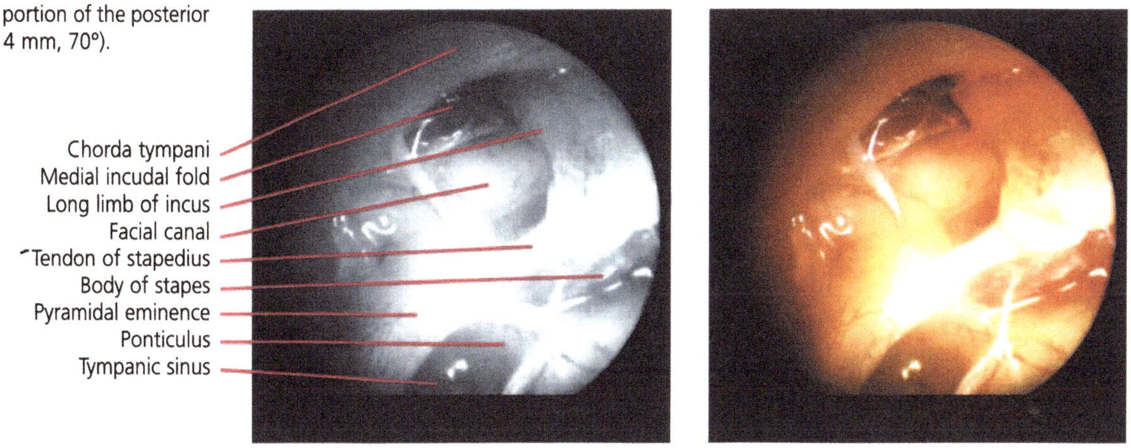

Fig. 26 Mid-segment of the posterior wall (rigid scope, 4 mm, 70°).

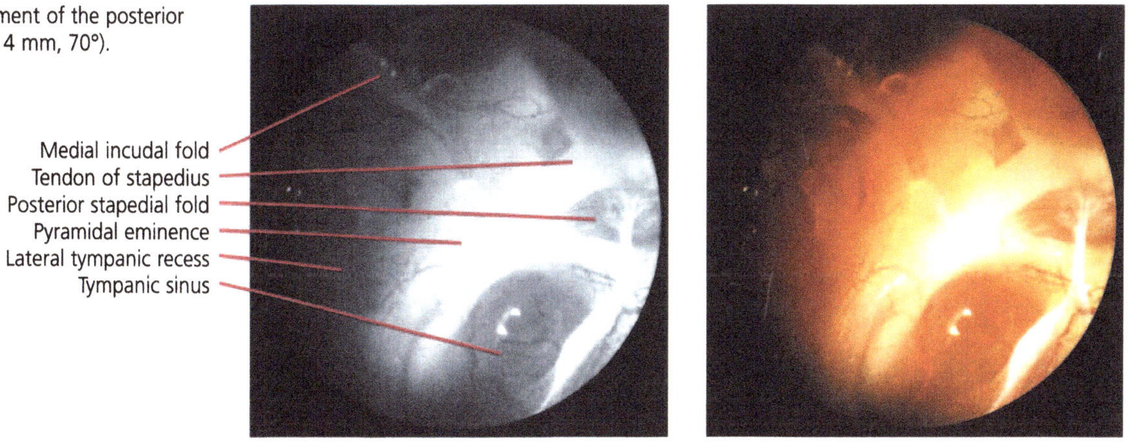

Fig. 27 Postero-inferior aspect of the posterior wall (rigid scope, 4 mm, 70°).

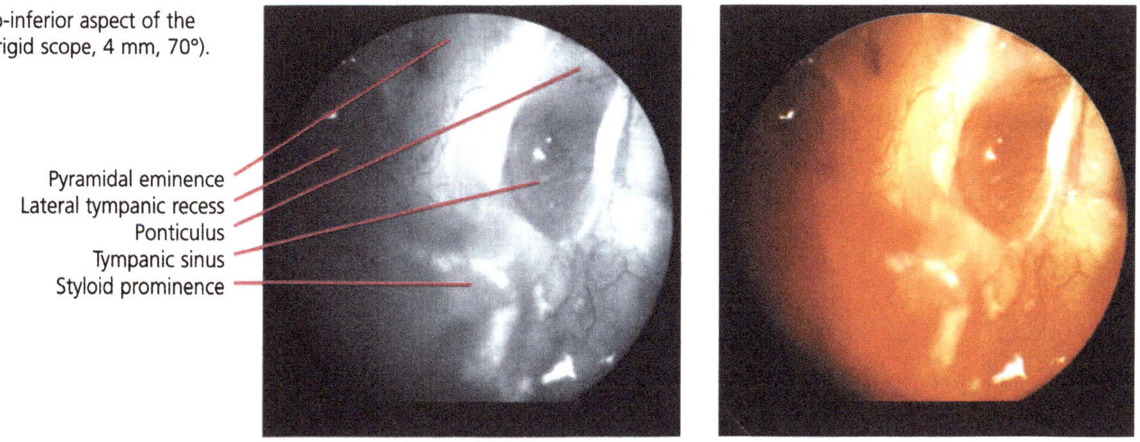

A 1.10 Exploration of the epitympanum

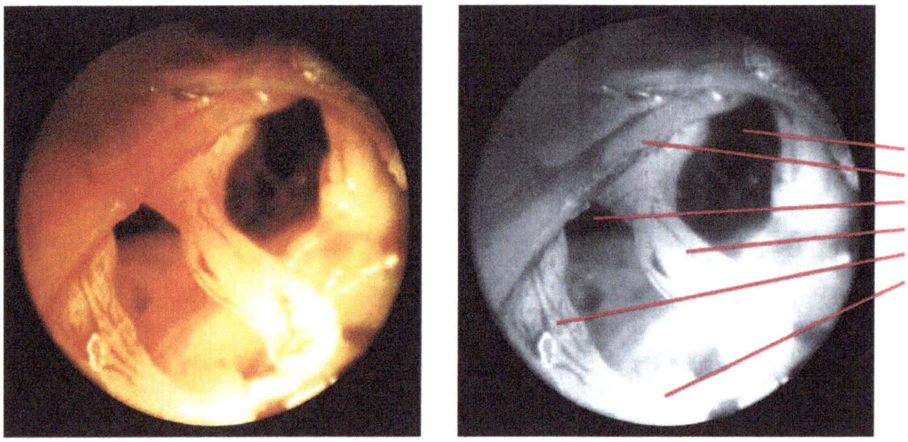

Fig. 28 View strictly upward with the chorda tympani seen from below (rigid scope, 4 mm, 70°, right ear).

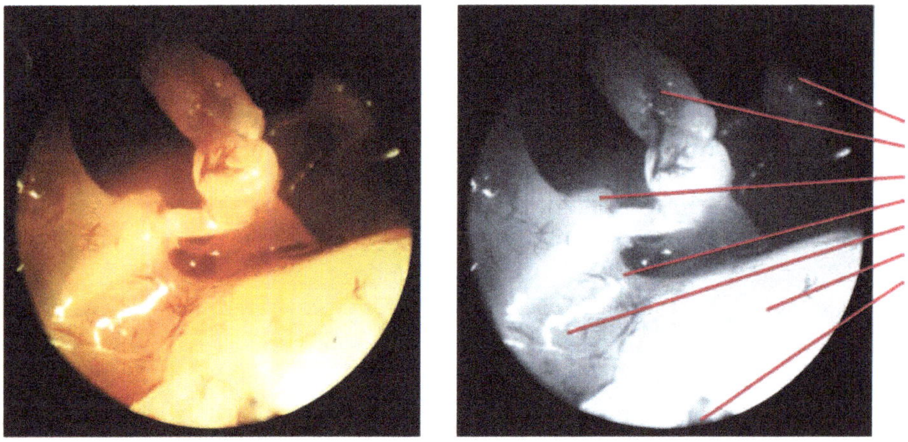

Fig. 29 Scope moved on 3 mm, lying on the bottom and looking up (rigid scope, 4 mm, 70°).

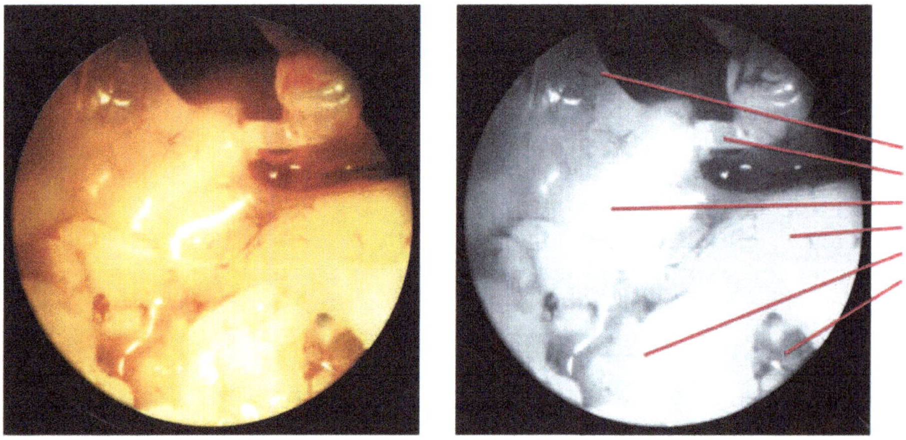

Fig. 30 Same view as in Fig 29, but scope advanced closer to round window niche (rigid scope, 4 mm, 70°).

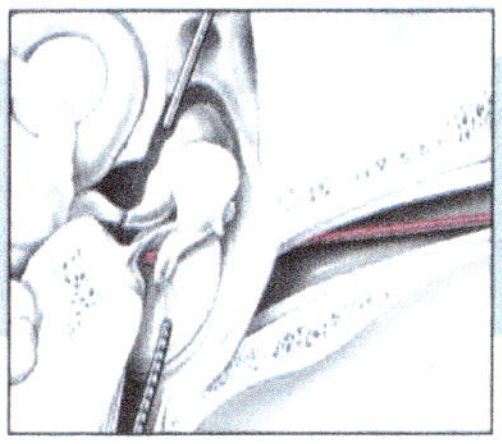

A 1.11 Hypotympanum

Seen by coaxial rotation of 30° scope

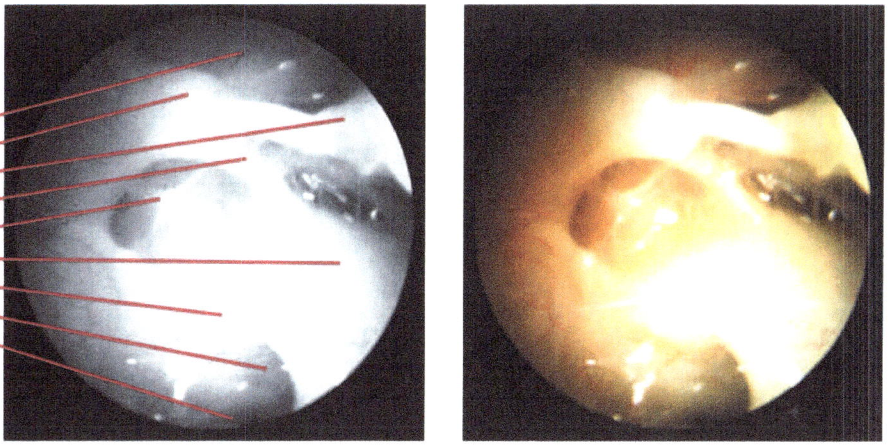

Fig. 31 Tympanic sinus and inferior part of posterior wall (rigid scope, 4 mm, 30°, right ear, looking posteriorly).

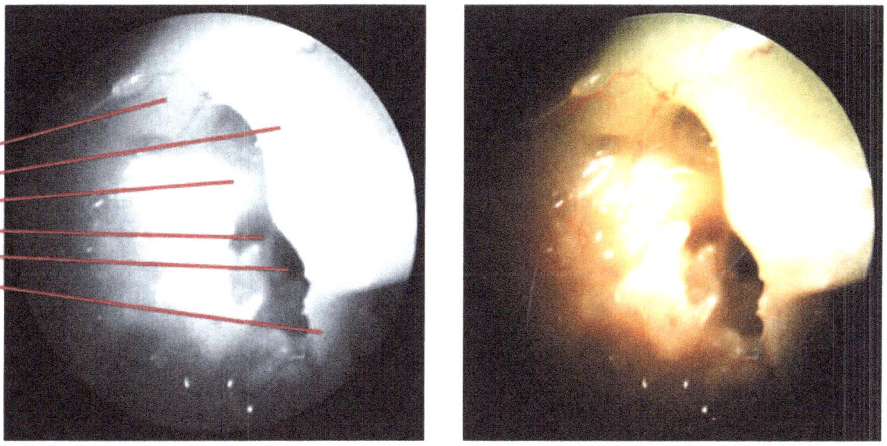

Fig. 32 Posterior part of the hypotympanum (rigid scope, 4 mm, 30°, looking down and posteriorly).

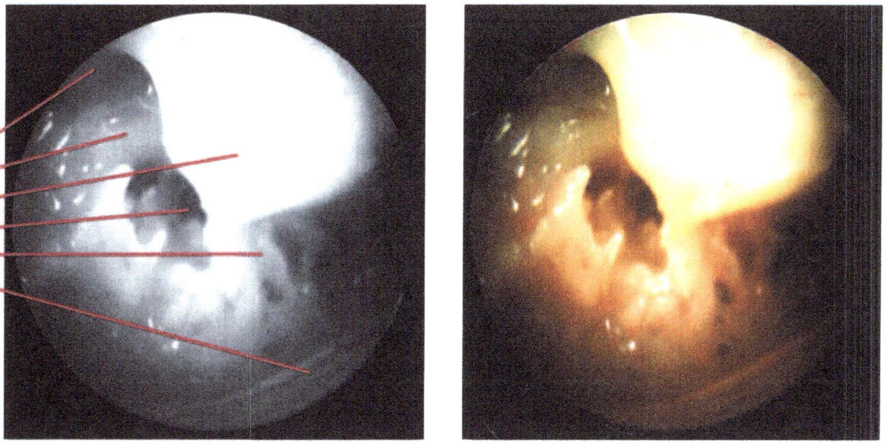

Fig. 33 Center of the hypotympanum (rigid scope, 4 mm, 30°, looking down).

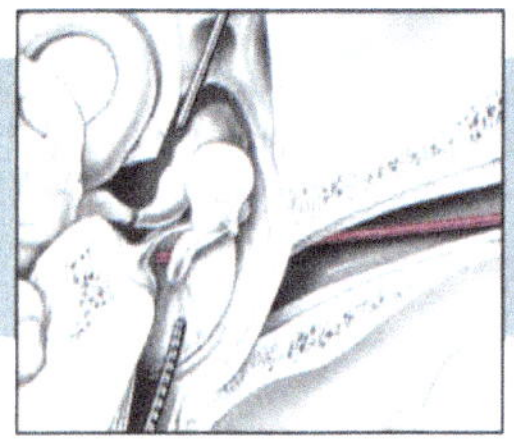

A 1.12 Some variants of the round window niche

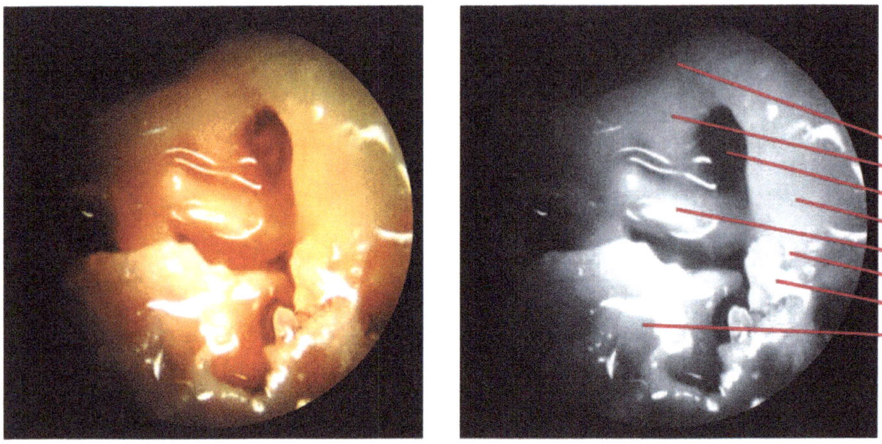

Fig. 34 Posterolateral opening of the niche with a large fustis (rigid scope, 4 mm, 30°).

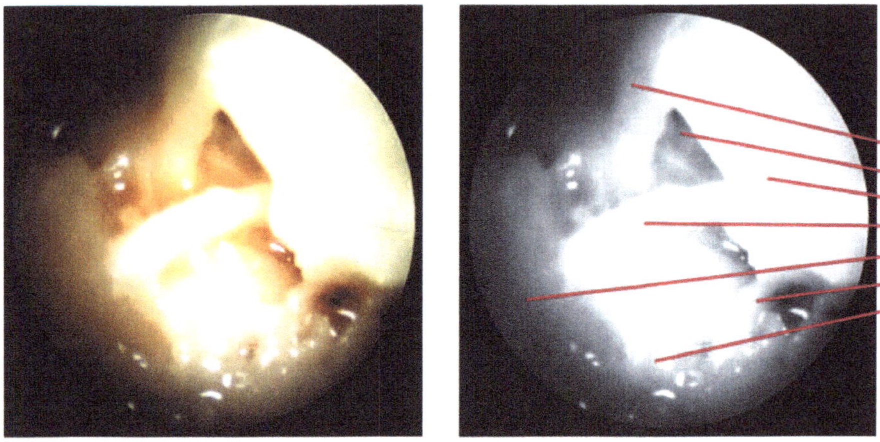

Fig. 35 Posterolateral opening of the niche with a narrow fustis (rigid scope, 4 mm, 30°).

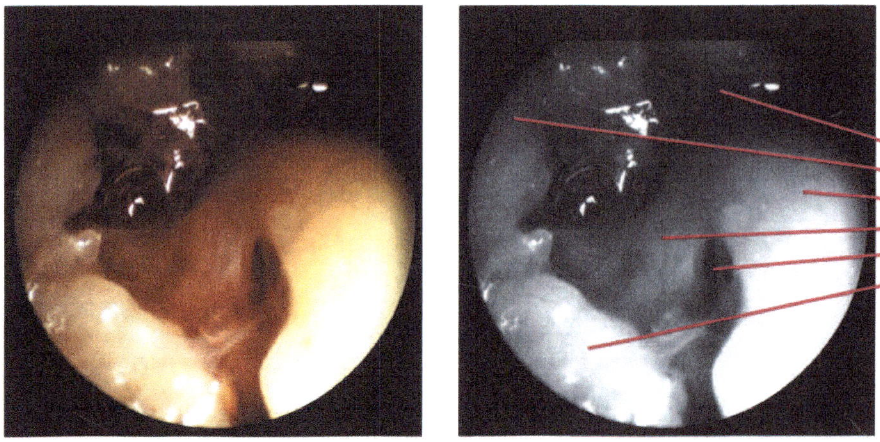

Fig. 36 Posterior opening of the niche without fustis (rigid scope, 4 mm, 70°).

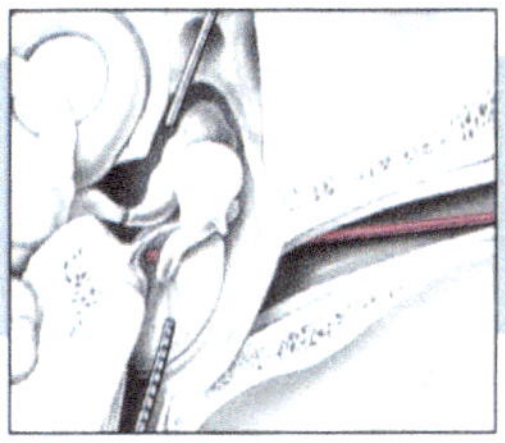

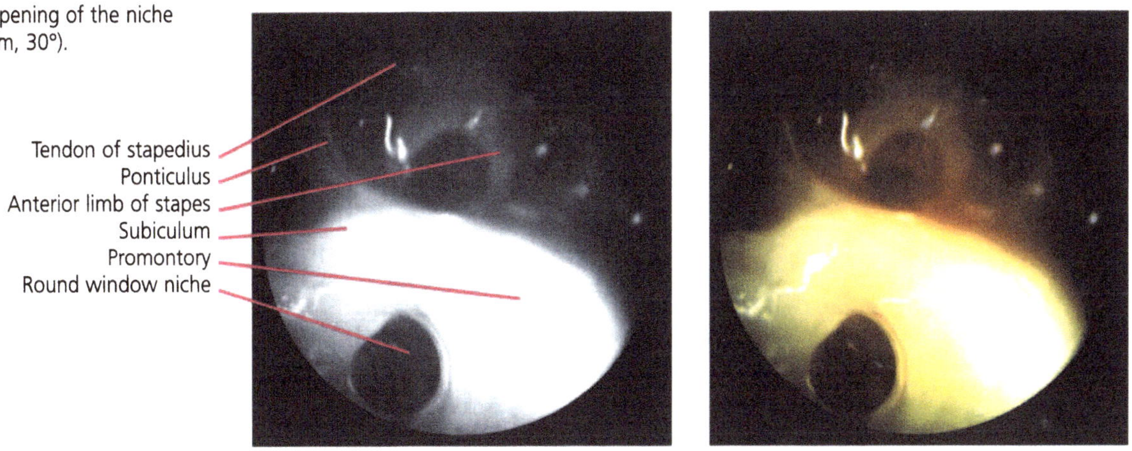

Fig. 37 Lateral opening of the niche (rigid scope, 4 mm, 30°).

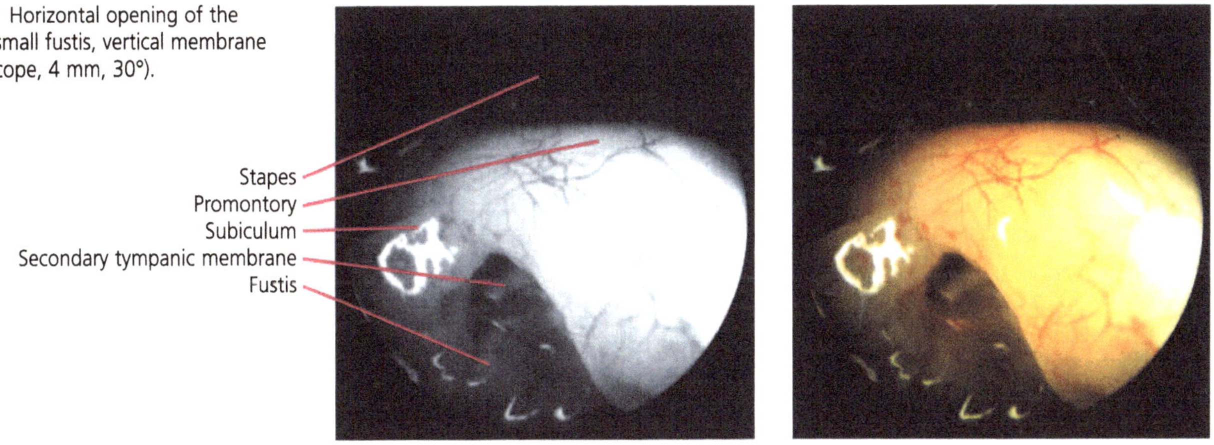

Fig. 38 Horizontal opening of the niche, small fustis, vertical membrane (rigid scope, 4 mm, 30°).

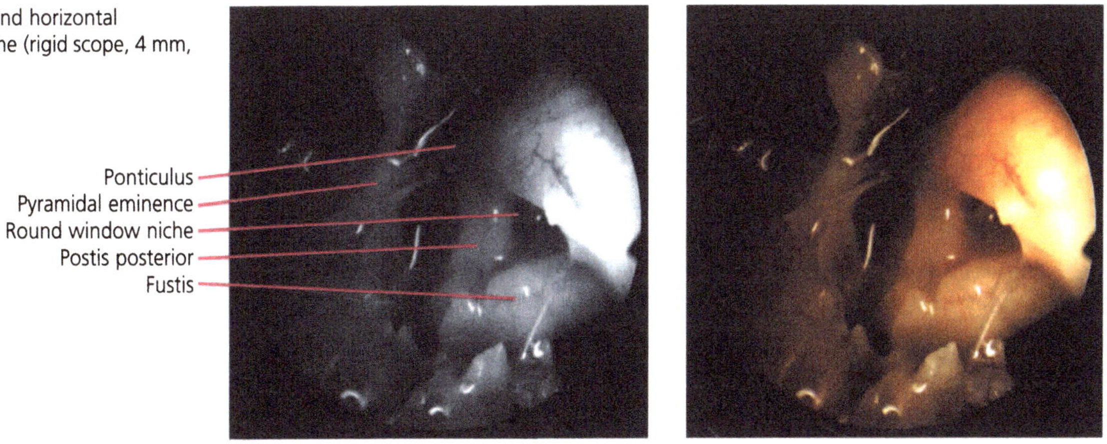

Fig. 39 Posterior and horizontal opening of the niche (rigid scope, 4 mm, 30°).

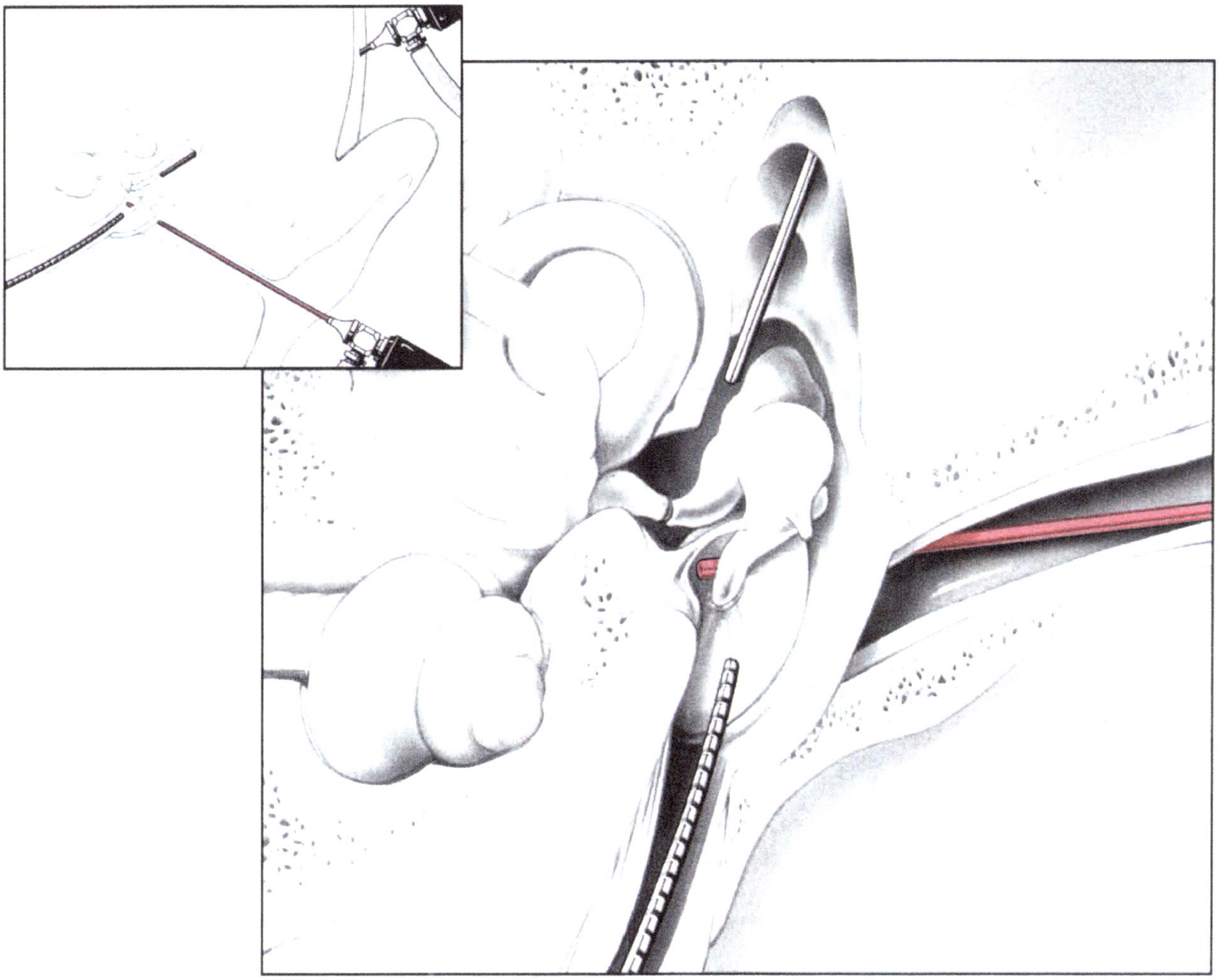

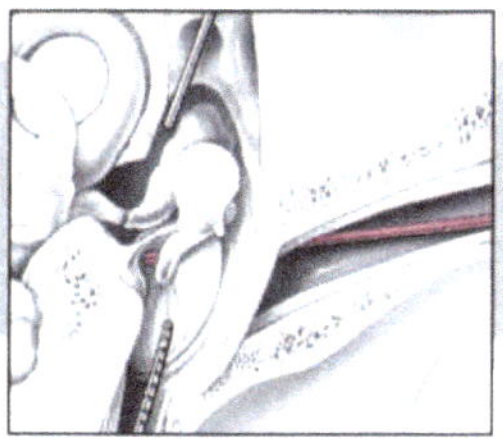

Transtympanic approach to the middle ear

The scope is introduced through a speculum and the tympanic membrane is incised under endoscopic vision either in the anterior inferior, the posterior superior or the posterior inferior quadrant. Incisions should always be radial from the tympanic ring towards the umbo, i.e. parallel to the arrangement of the fibers in the stratum fibrosum of the tympanic membrane so that, when the scope is advanced beyond the incision, the fibers are separated, but preserved.

The site of the paracentesis, i.e. the quadrant to be incised, depends on the structures of interest. For this procedure rigid angled 30° and 60° scopes with an outer diameter of 1.9 mm or 2.7 mm are used.

(1) Incisions in the anterior inferior quadrant give an excellent overview of the protympanum. Rotating angled scopes (30° or 60°) in the long axis provides an extended field of view.

(2) Through incisions in the posterior superior quadrant 30° scopes will show the entire posterior wall from the chorda tympani to the oval fossa. With a 60° scope the area from the anterior epitympanum to the aditus ad antrum is seen when looking up and the protympanum and hypotympanum when looking down. (In this approach care should be taken to spare the incudostapedial joint.)

(3) 30° scopes inserted through an incision in the posterior inferior quadrant bring into view the round window niche, the stapes and the posterior wall of the tympanic cavity. Like in (2) rotation of a 60° scope shows the entire region from the roof of the protympanum across the medial wall to the bottom. [Note that the jugular dome may be high and that there may be dehiscences in the bone.]

Anterior superior quadrant paracentesis does not offer any additional benefits.

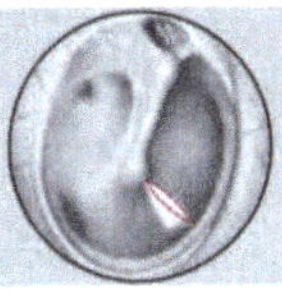

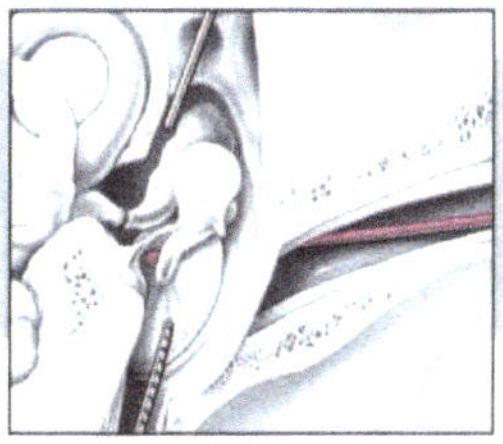

B 1 Anterior inferior transtympanic approach

B 1.1 Site of approach

Fig. 40 Radial incision in the anterior inferior quadrant of a right tympanic membrane (rigid scope, 4 mm, 0°, right ear).

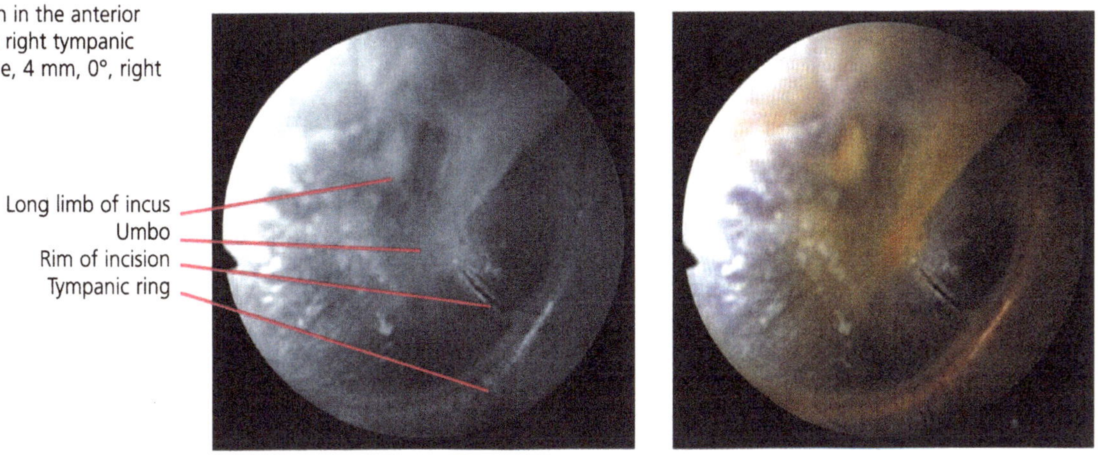

Fig. 41 Monitored through a rigid 4 mm 0° scope the 2.7 mm scope is introduced.

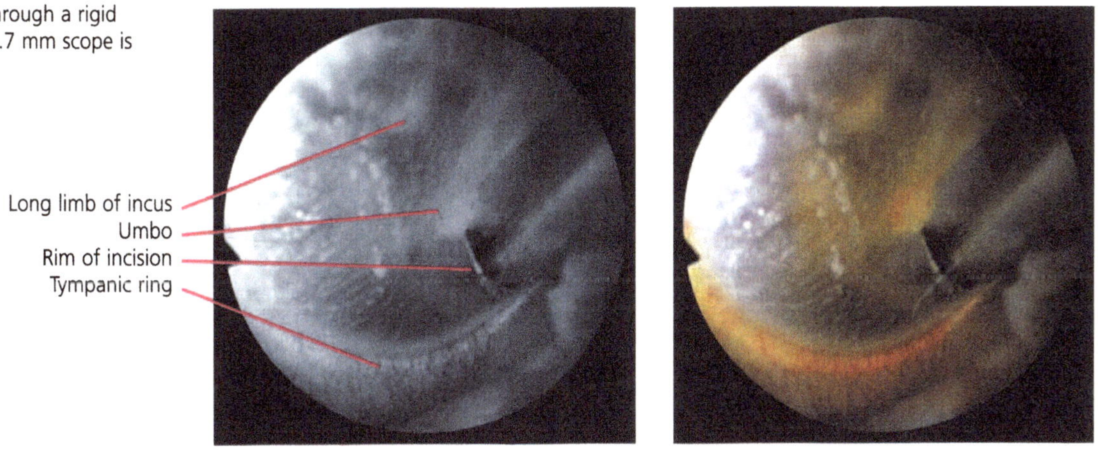

Fig. 42 Tympanic cavity viewed through the incision (rigid scope, 2.7 mm, 30°).

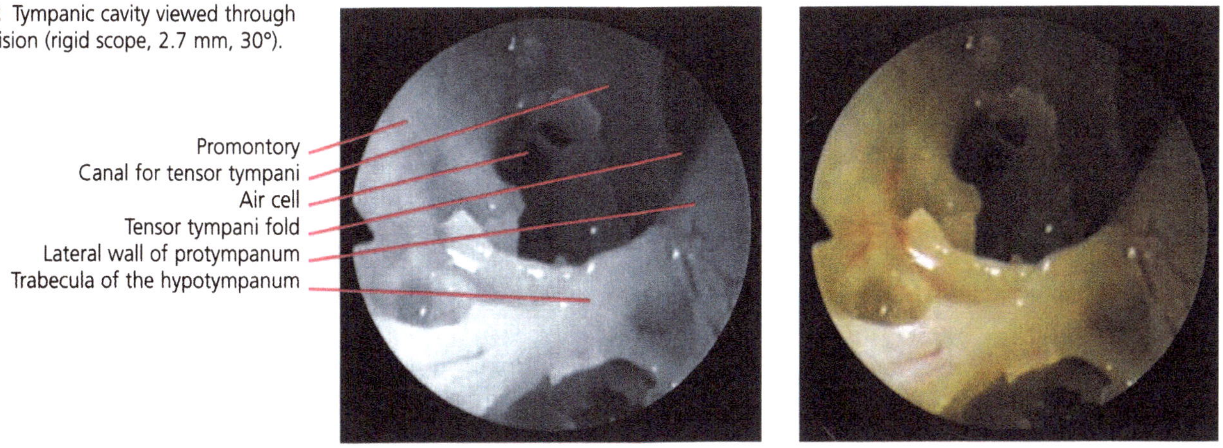

B 1.2 Overview of protympanum from the top down

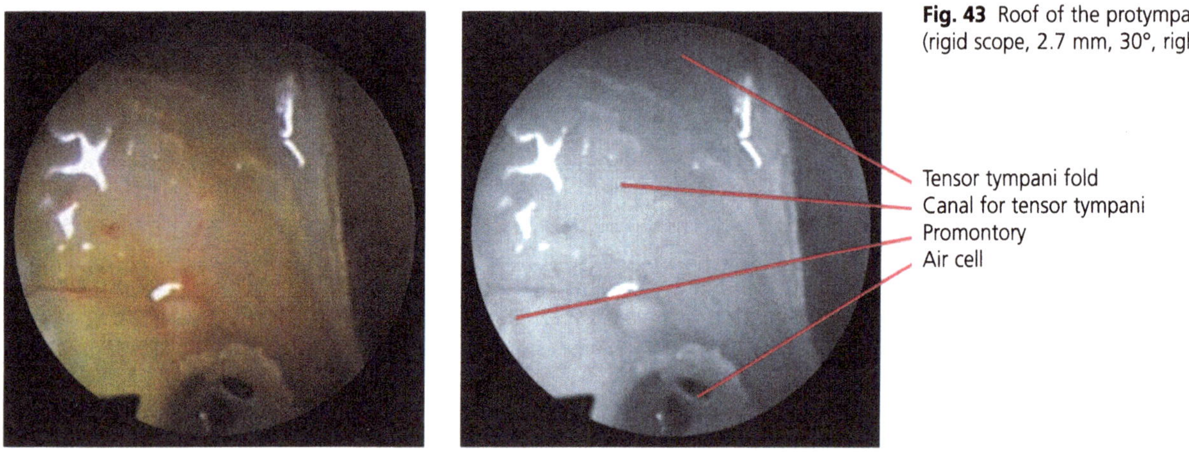

Fig. 43 Roof of the protympanum (rigid scope, 2.7 mm, 30°, right ear).

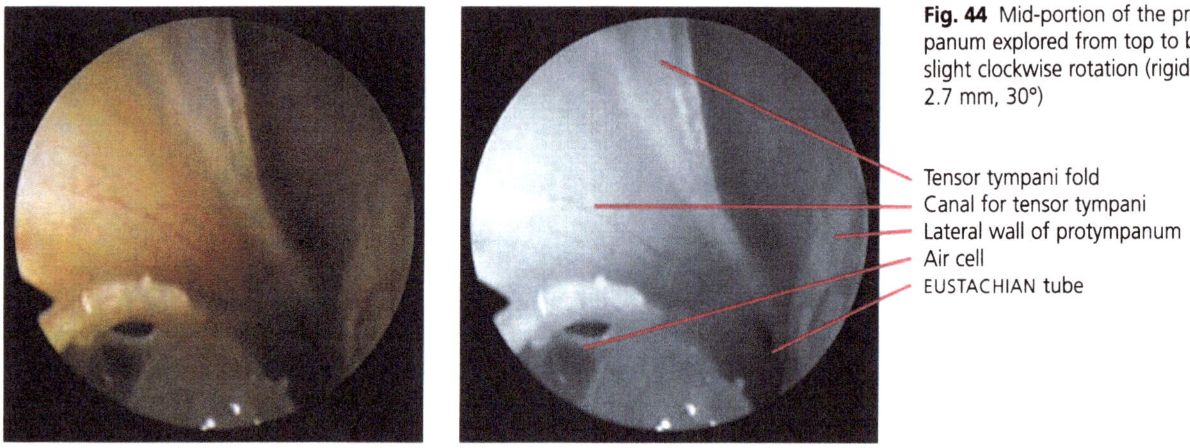

Fig. 44 Mid-portion of the protympanum explored from top to bottom by slight clockwise rotation (rigid scope, 2.7 mm, 30°)

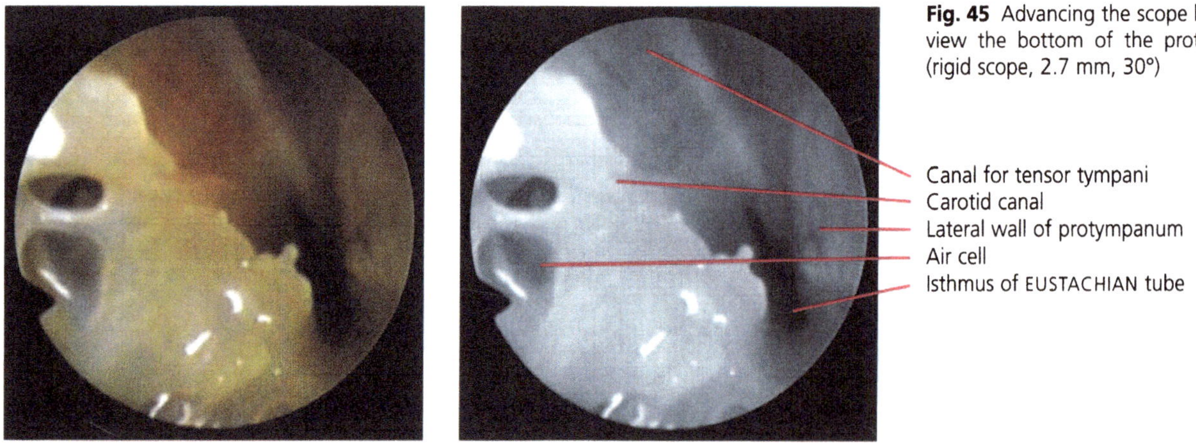

Fig. 45 Advancing the scope brings into view the bottom of the protympanum (rigid scope, 2.7 mm, 30°)

B 1.3.1 Protympanic region close to tympanic membrane

Fig. 46 The scope approaches the incision in the tympanic membrane (rigid scope, 1.9 mm, 30°, looking forward, right ear).

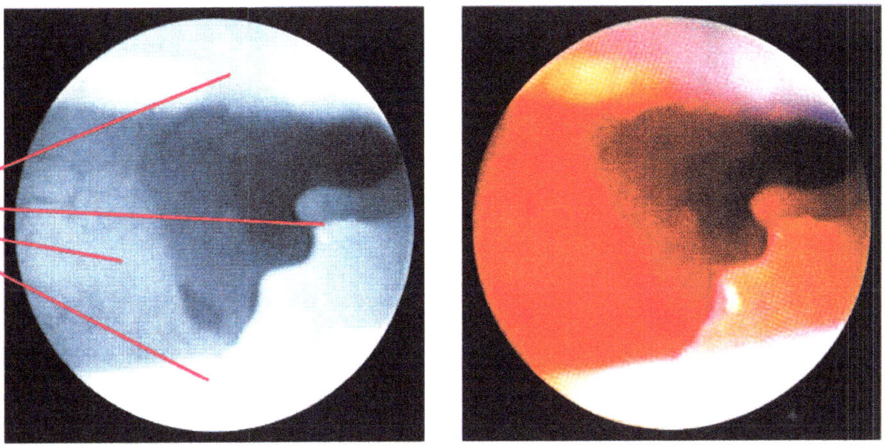

Fig. 47 The scope lies in the plane of the tympanic incision and looks at the protympanum (rigid scope, 1.9 mm, 30°).

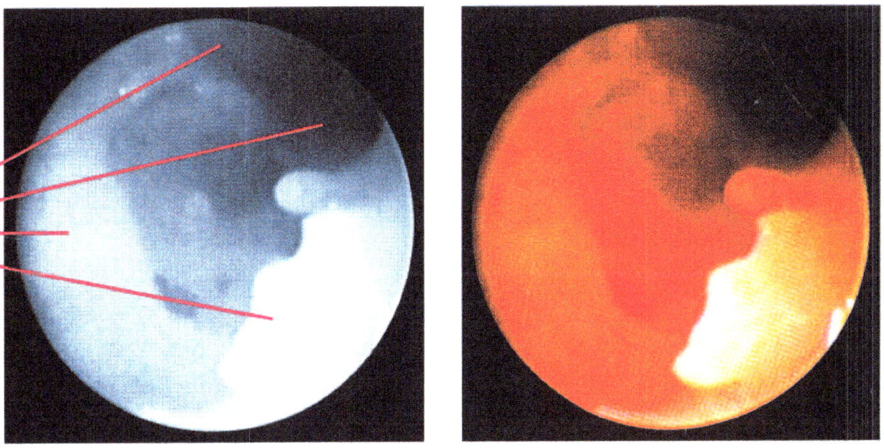

Fig. 48 Overview of the protympanum after advancing the scope (rigid scope, 1.9 mm, 30°).

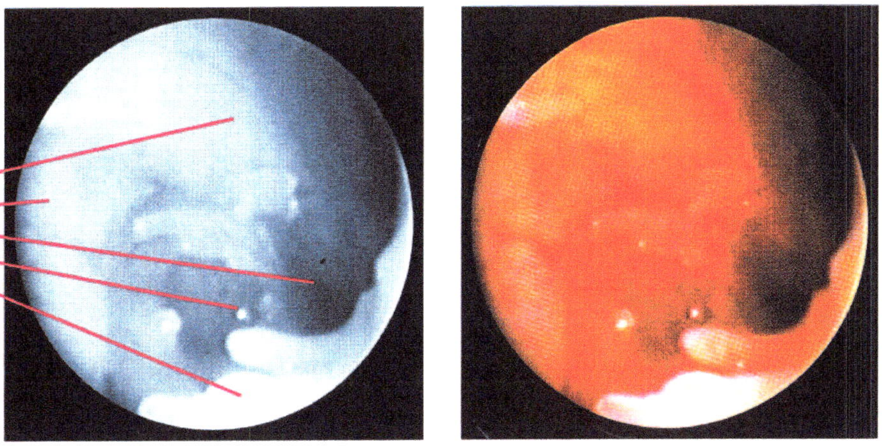

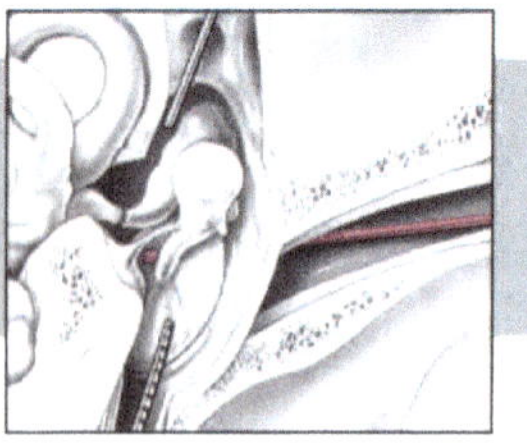

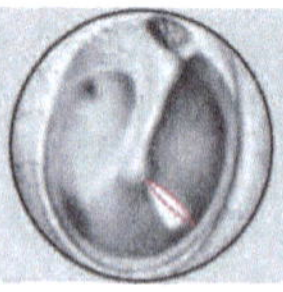

B 1.3.2 Medial wall of protympanum and tympanic opening of the EUSTACHIAN tube

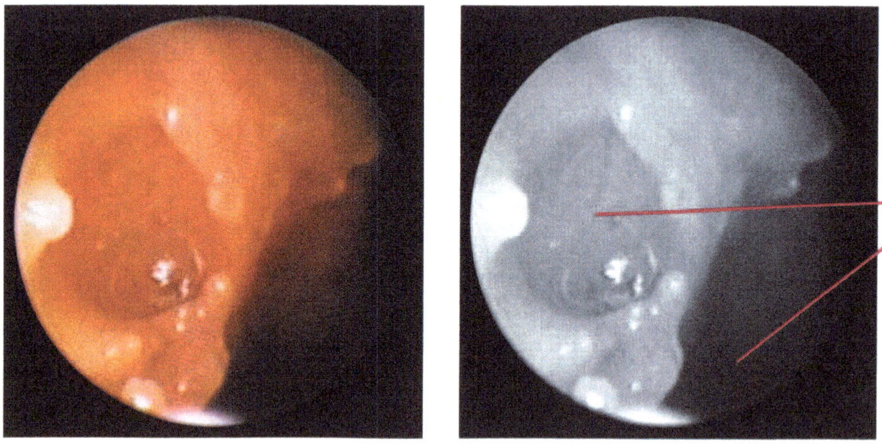

Fig. 49 Scope focusing on an air cell in the medial protympanic wall (do not confuse with EUSTACHIAN tube opening), (rigid scope, 1.9 mm, 30°, right ear).

Air cell
EUSTACHIAN tube

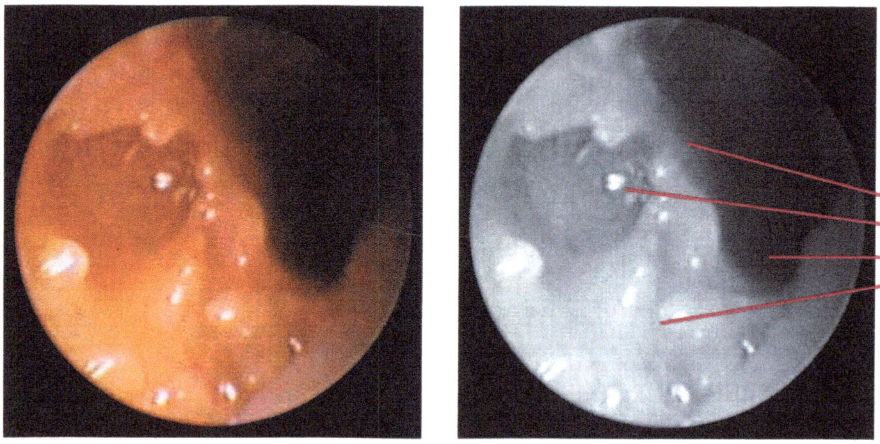

Fig. 50 Anterior inferior view of the bottom of the protympanum (rigid scope, 1.9 mm, 30°).

Prominence of carotid canal
Air cell
Tympanic opening of EUSTACHIAN tube
Bottom of protympanum

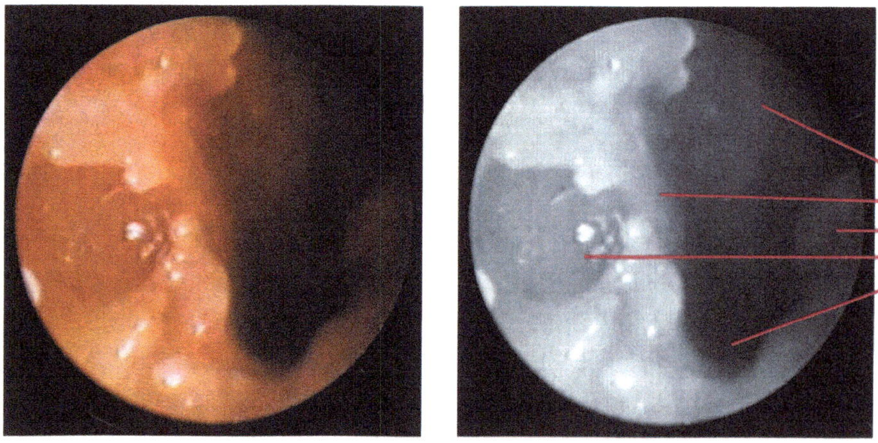

Fig. 51 Anterior superior view of the roof of the protympanum (rigid scope, 1.9 mm 30°).

Canal of tensor tympani
Prominence of carotid canal
Lat. wall of protympanum
Air cell
Tympanic opening of EUSTACHIAN tube

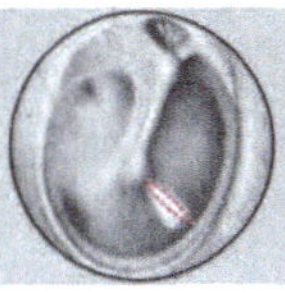

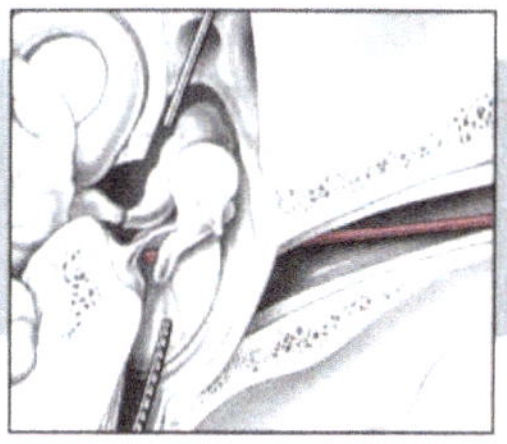

B 1.4.1 Center of protympanum

Fig. 52 Overview of the protympanum seen with the scope looking forward (rigid scope, 1.9 mm, 60°, right ear).

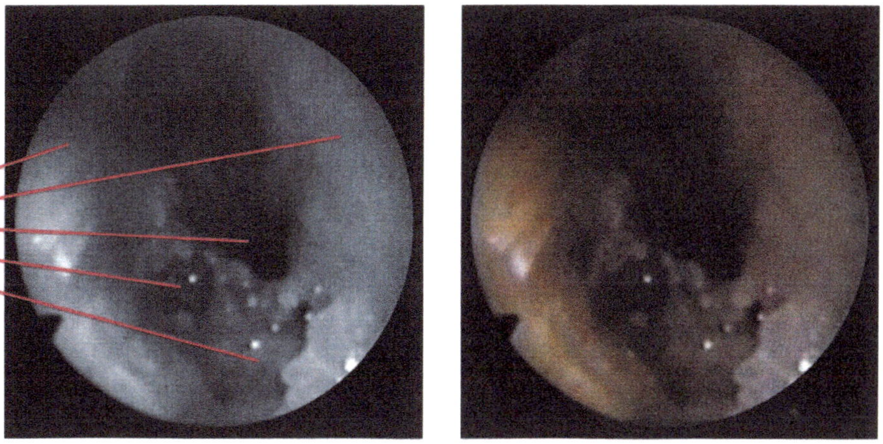

Fig. 53 Scope looking forward in a more advanced position (rigid scope, 1.9 mm, 60°).

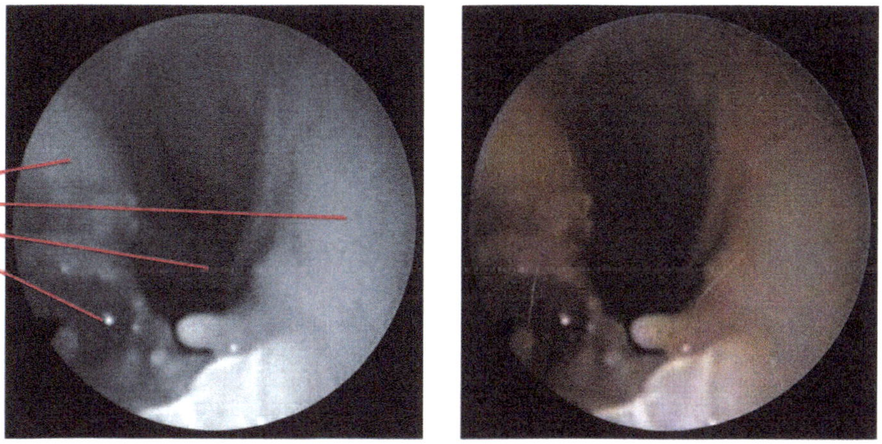

Fig. 54 Scope slightly rotated clockwise to view the isthmus of the EUSTACHIAN tube (rigid scope, 1.9 mm, 60°).

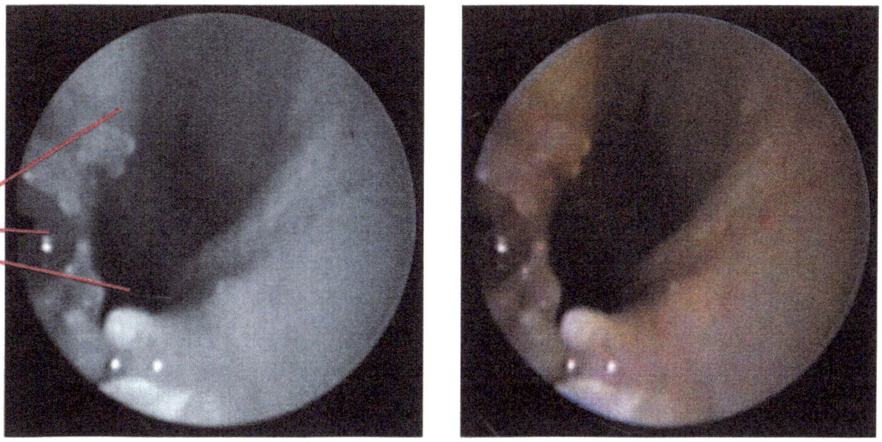

B 1.4.2 Roof of protympanum

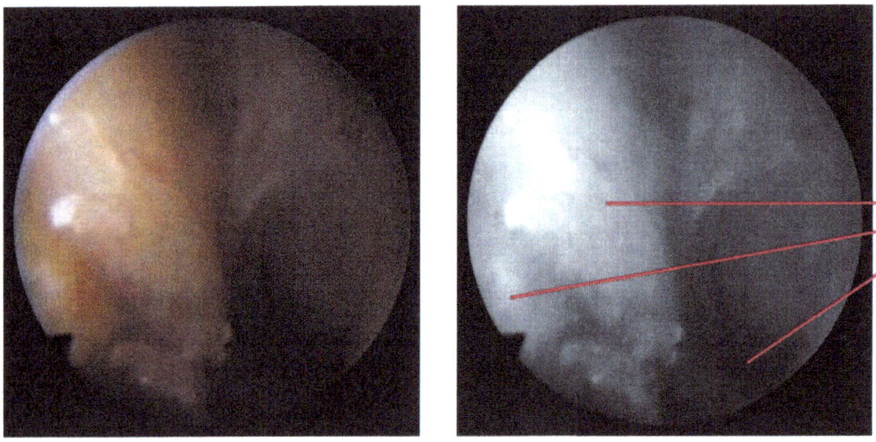

Fig. 55 Scope looking forward and up to focus on the prominence of the canal for the tensor tympani (rigid scope, 1.9 mm, 60°, right ear).

Canal of tensor tympani
Carotid canal
Tympanic opening of EUSTACHIAN tube

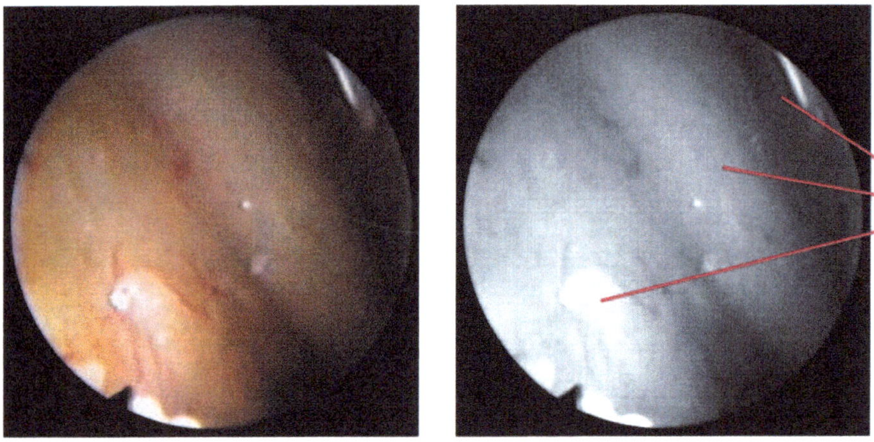

Fig. 56 Scope rotated counterclockwise to extend upward view (rigid scope, 1.9 mm, 60°).

Mucosal fold of tensor tympani
Canal of tensor tympani
Carotid canal

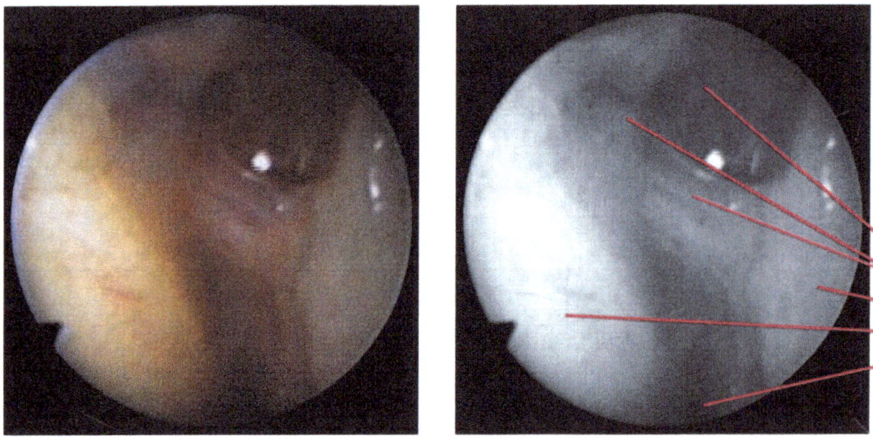

Fig. 57 The tip of the scope is lifted and still looks upwards: Through the translucent mucosal fold of the tensor tympani the anterior epitympanic recess can be seen. In this case there is a communication with the protympanum through a dehiscence in the mucosal fold (= supratubal recess), (rigid scope, 1.9 mm, 30°).

Anterior epitympanic recess
Mucosal fold of tensor tympani
Lateral wall of protympanum
Canal of tensor tympani
Tympanic opening of EUSTACHIAN tube

B 1.4.3 Bottom of protympanum

Fig. 58 Scout view of the protympanum with the scope looking forward (compare Fig. 13), (rigid scope, 1.9 mm, 60°, right ear).

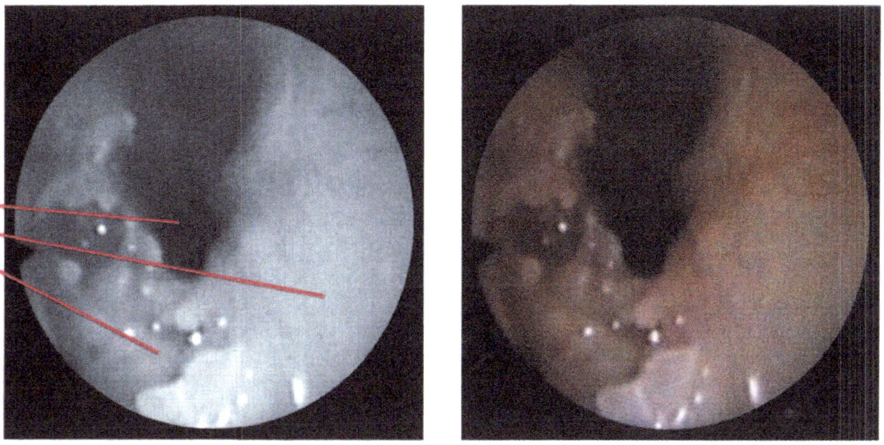

Fig. 59 Scope rotated clockwise and directed forward and downward (rigid scope, 1.9 mm, 60°).

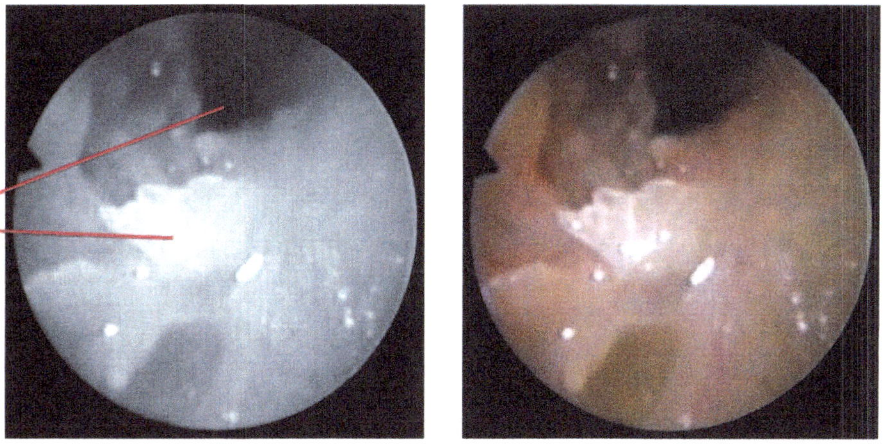

Fig. 60 Scope retracted to expose the hypotympanum in the plane of the anterior tympanic membrane margin (rigid scope, 1.9 mm, 60°).

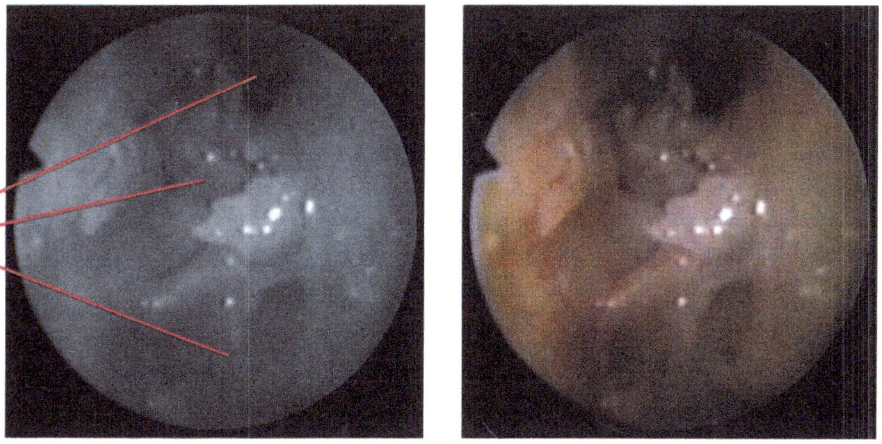

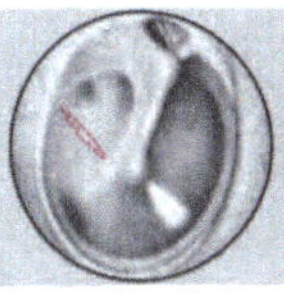

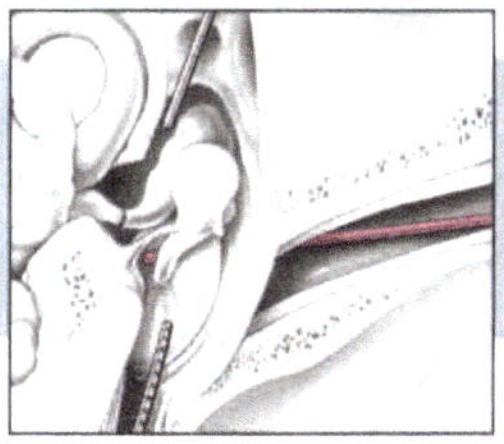

B 2 Posterior superior transtympanic approach

1.9 mm rigid scope

B 2.1 Site of approach

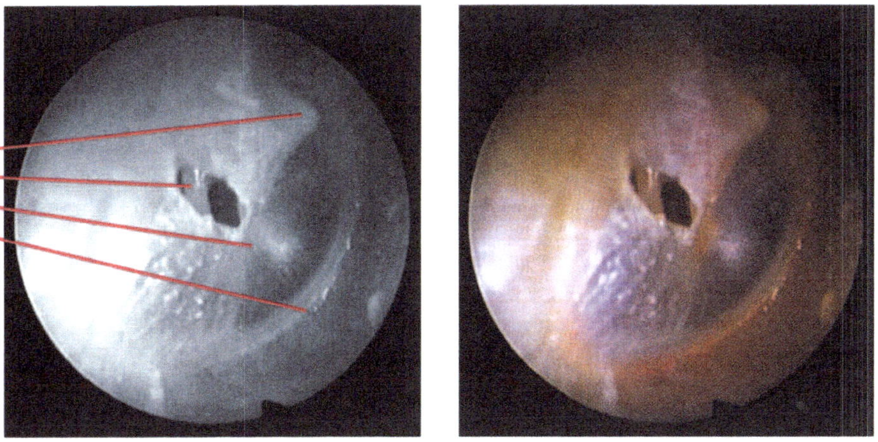

Fig. 61 Incision in the posterior superior quadrant of a right tympanic membrane (rigid scope, 4 mm, 0°, right ear).

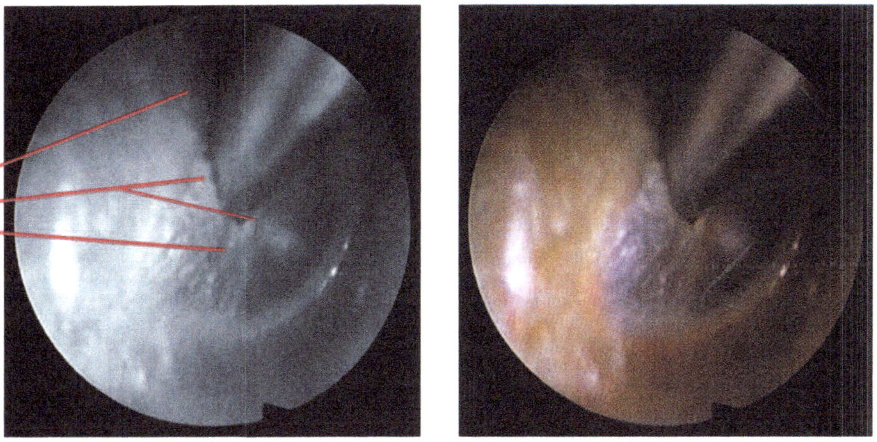

Fig. 62 Monitored through a rigid 4 mm 0° scope, the 1.9 mm rigid scope is introduced.

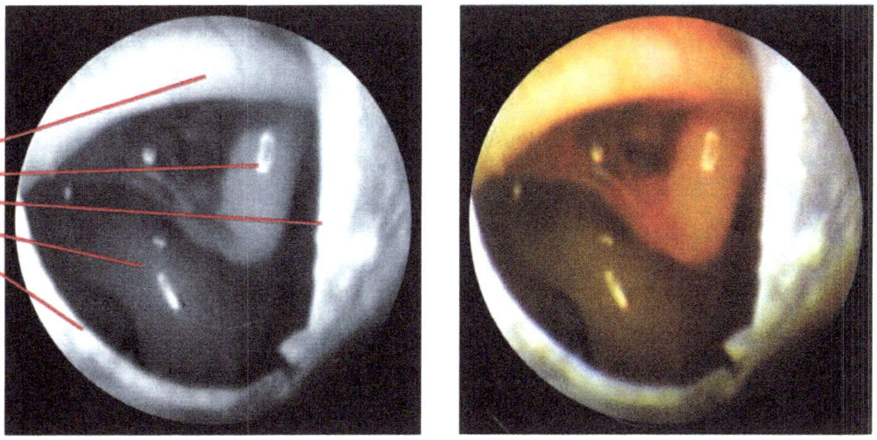

Fig. 63 Tympanic cavity viewed through the incision (rigid scope, 1.9 mm, 0°, wide angle).

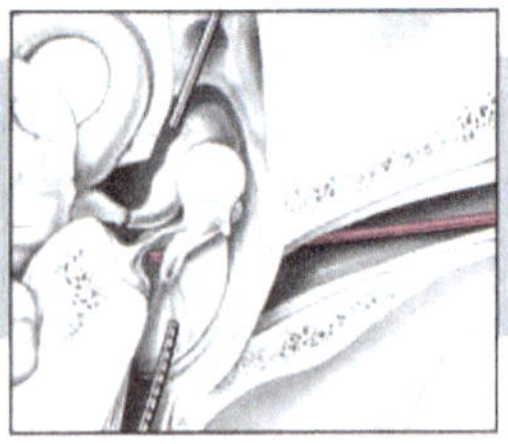

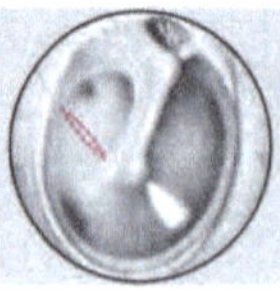

B 2.2.1 Course of chorda tympani

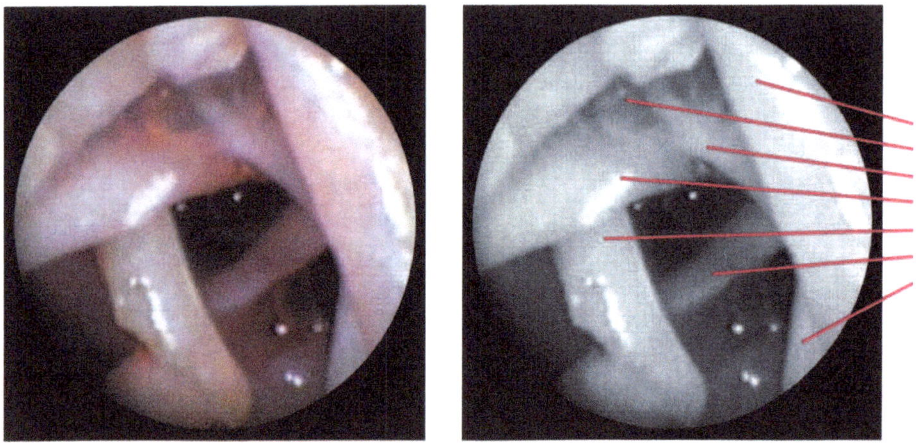

Fig. 64 Overview in a superior direction (rigid scope, 1.9 mm, 30°, right ear).

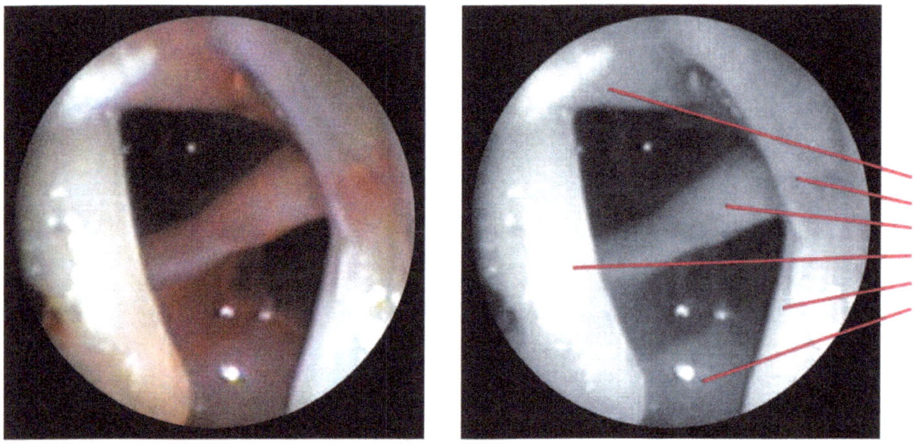

Fig. 65 Superior anterior view: chorda tympani crossing tendon of tensor tympani (rigid scope, 1.9 mm, 30°).

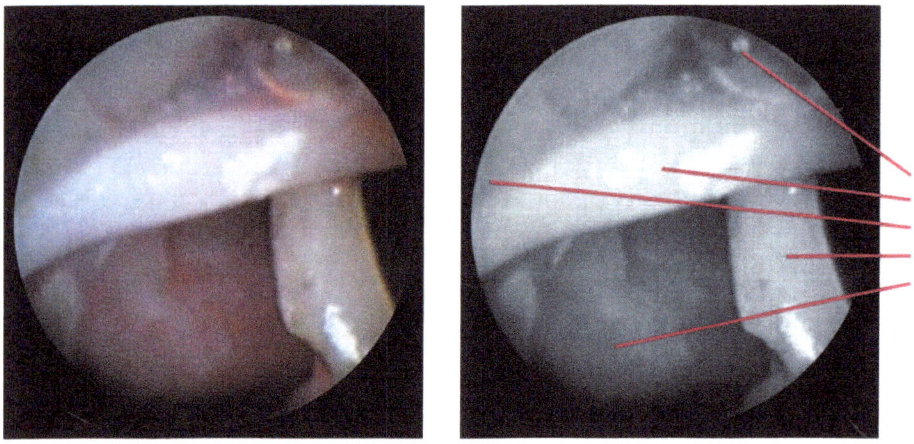

Fig. 66 Superior posterior view: emergence of chorda tympani at chordal eminence (rigid scope, 1.9 mm, 30°).

B 2.2.2 Posterior wall with pyramidal eminence

Fig. 67 Scope looks posteriorly and up to follow the chorda tympani to the chordal eminence (rigid scope, 1.9 mm, 30°).

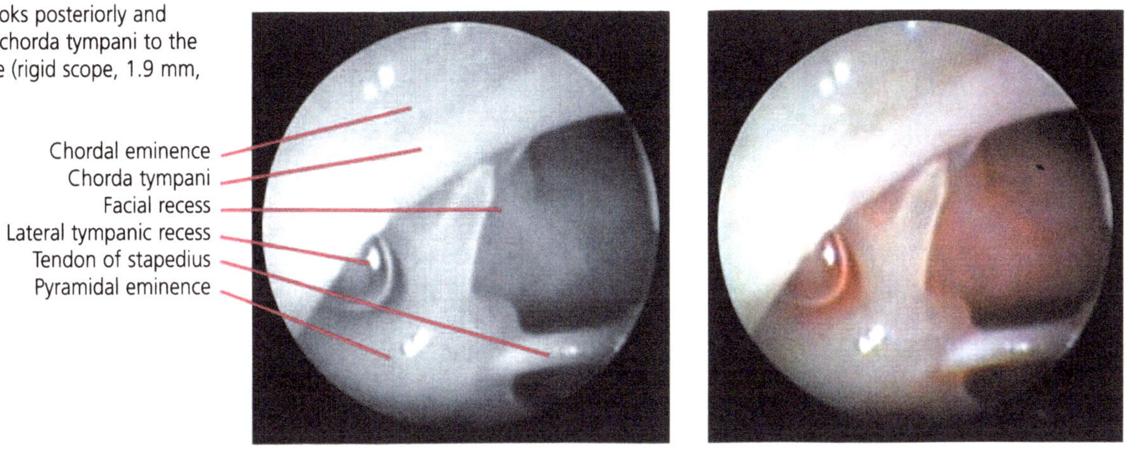

Fig. 68 View in a strictly posterior direction with the pyramidal eminence in the center (rigid scope, 1.9 mm, 30°).

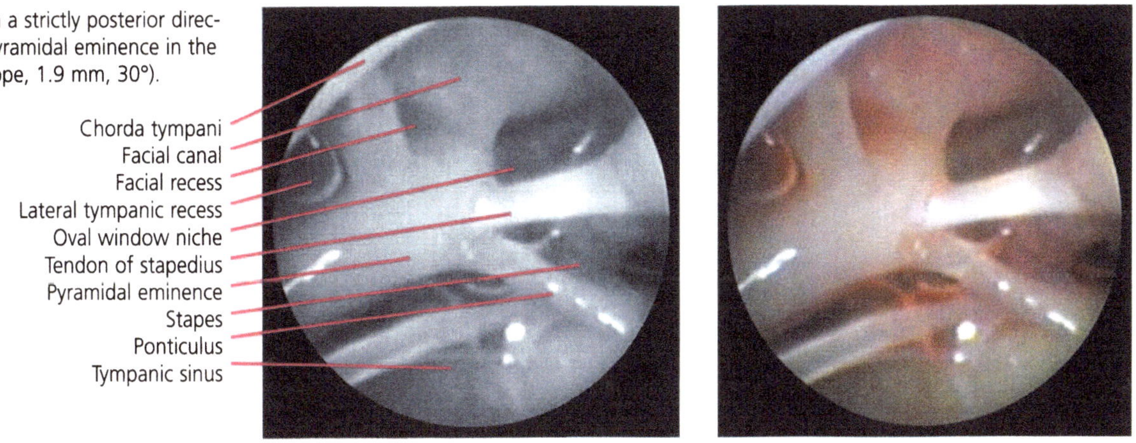

Fig. 69 Scope looks posteriorly and down at the structures below the pyramidal eminence (rigid scope, 1.9 mm, 30°).

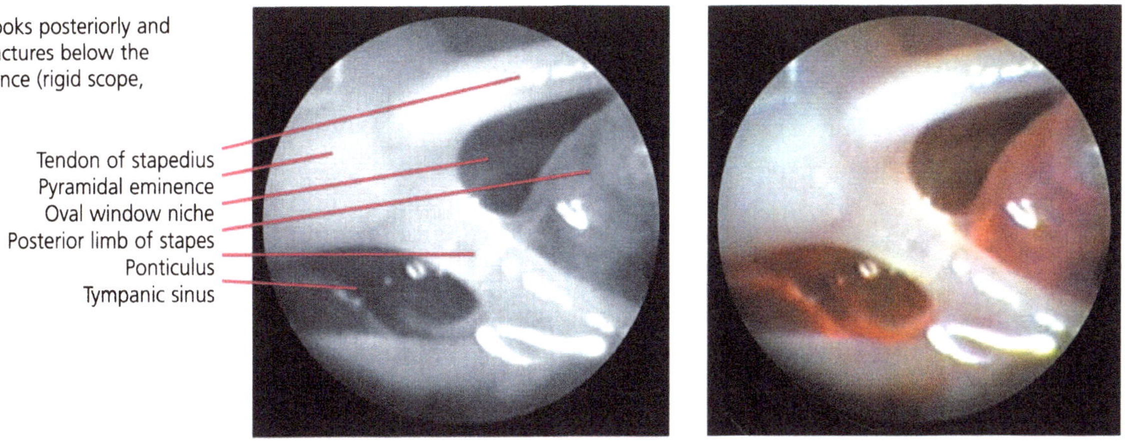

B 2.2.3 Facial canal

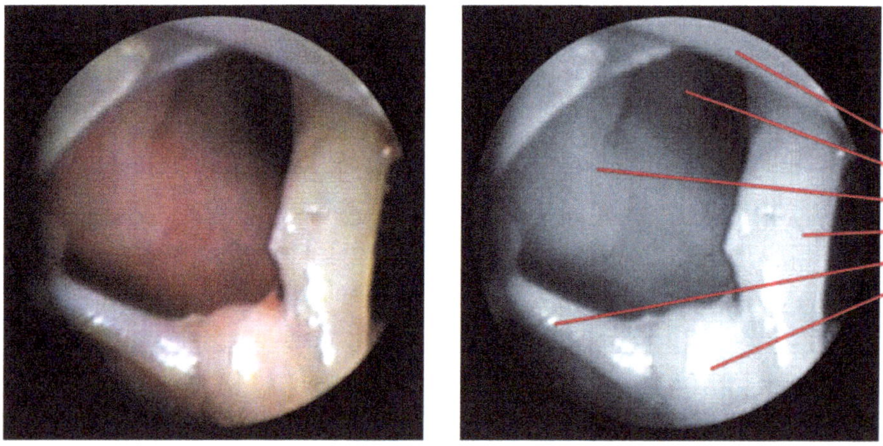

Fig. 70 Overview: Tip of the scope in the plane of the chorda tympani (rigid scope, 1.9 mm, 30°, right ear).

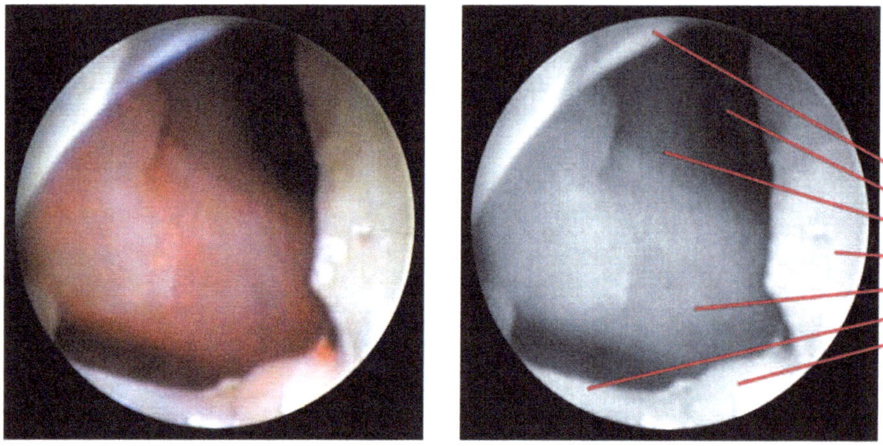

Fig. 71 Closer view of the facial canal through a triangle formed by the long limb of incus, the tendon of stapedius and the mucosal fold of the chorda tympani (rigid scope, 1.9 mm, 30°).

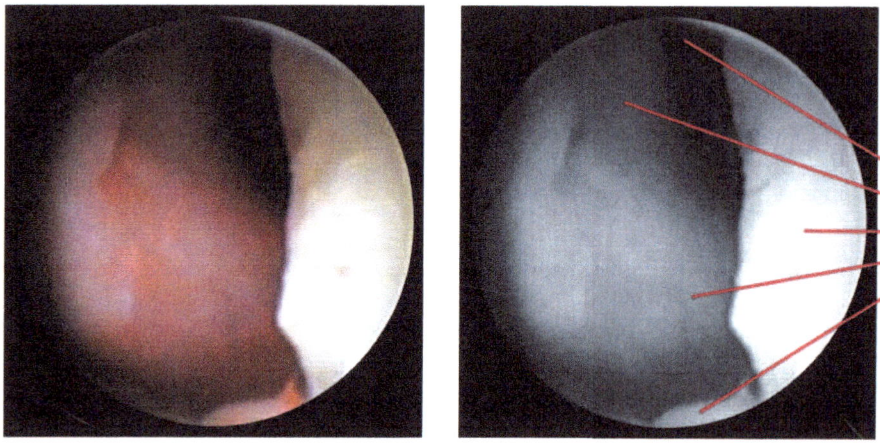

Fig. 72 Scope further advanced above the tendon of stapedius (2 mm in front of facial canal), (rigid scope, 1.9 mm 30°).

B 2.3.1 Oval window niche

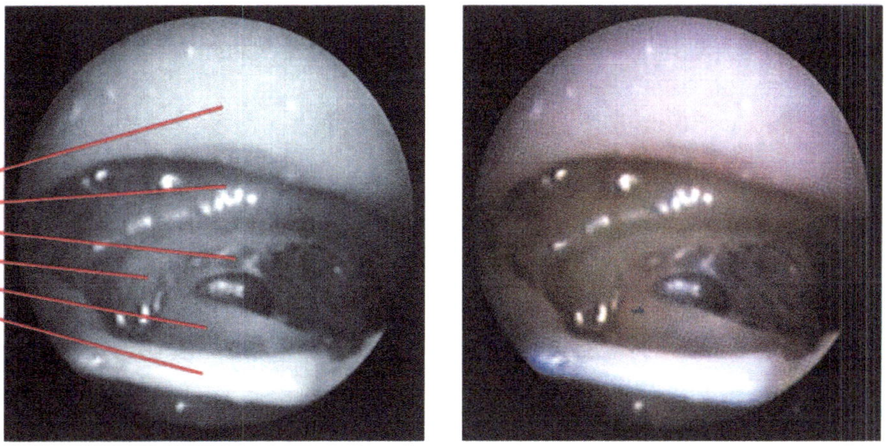

Fig. 73 Superior posterior margin of the oval fossa (rigid scope, 1.9 mm, 30°, right ear).

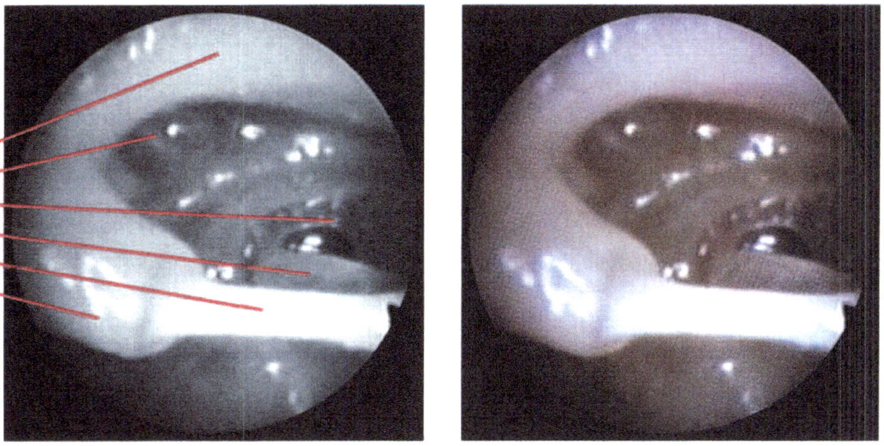

Fig. 74 Posterior boundaries of oval fossa (rigid scope, 1.9 mm, 30°).

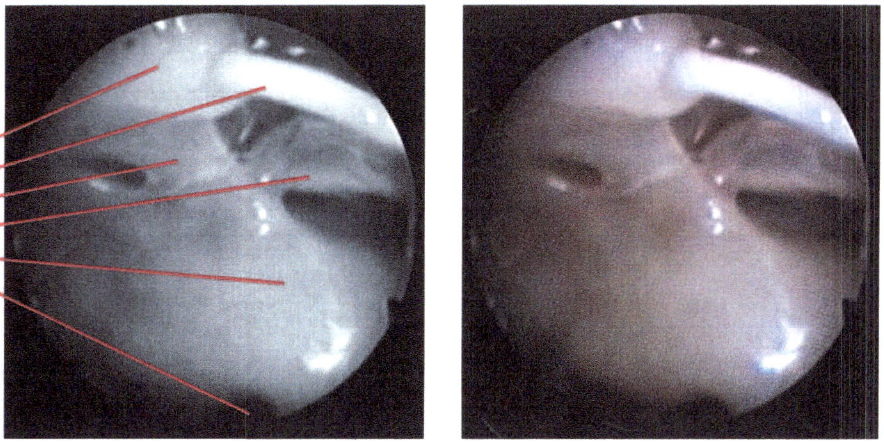

Fig. 75 Posterior inferior boundaries of oval window niche (rigid scope, 1.9 mm, 30°).

B 2.3.2 Stapes

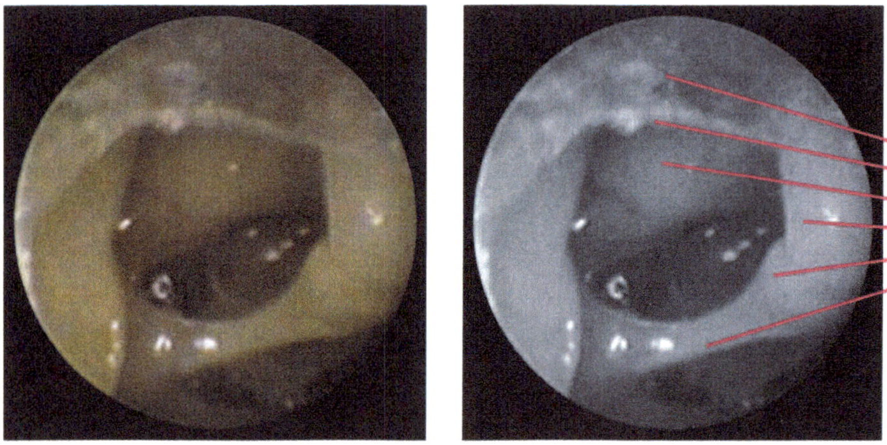

Fig. 76 Scope in the plane of the incision (rigid scope, 1.9 mm, 30°, right ear).

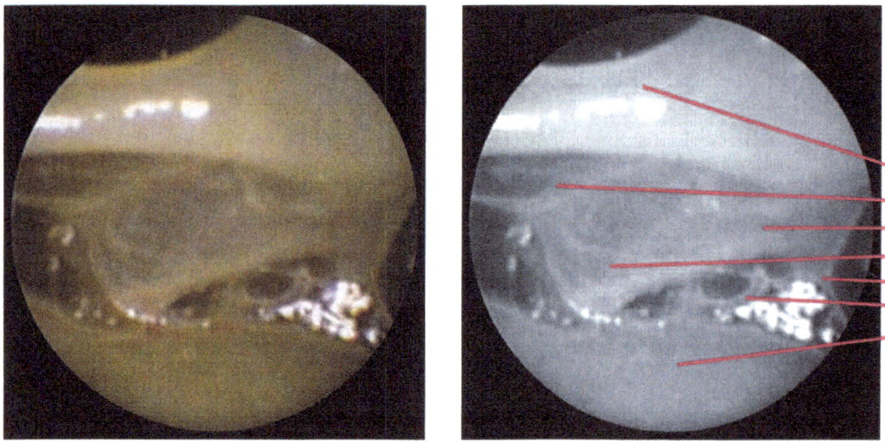

Fig. 77 Scope slightly tipped downward looking at the insertion of the stapedius muscle (rigid scope, 1.9 mm, 30°).

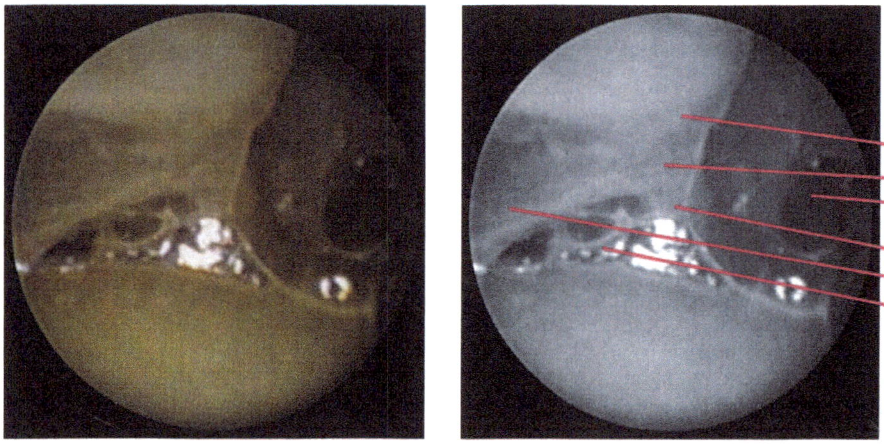

Fig. 78 View of the anterior inferior boundaries of the oval window niche (rigid scope, 1.9 mm, 30°).

B 2.3.3 Anterior tympanic isthmus

Fig. 79 Scope in the plane of the chorda tympani looks forward and up and focuses on the cochlearifrom process (rigid scope, 1.9 mm, 30°, right ear).

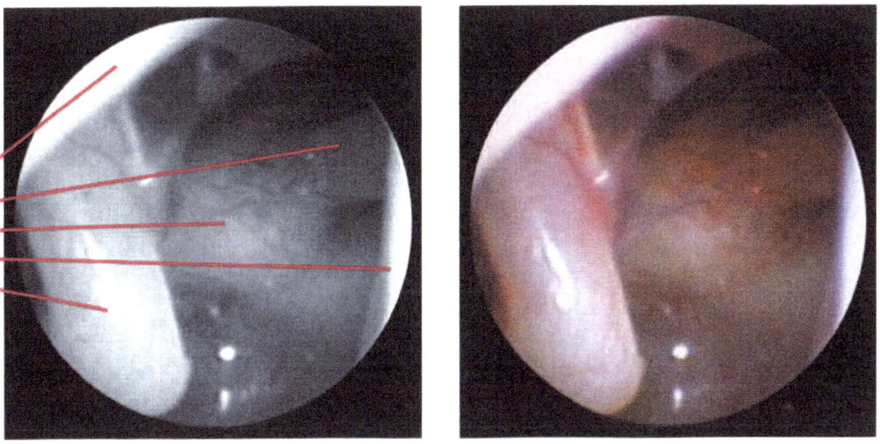

Fig. 80 Scope closer to cochleariform process than in Fig. 79. Note membrane partly obturating anterior tympanic isthmus (rigid scope, 1.9 mm, 30°).

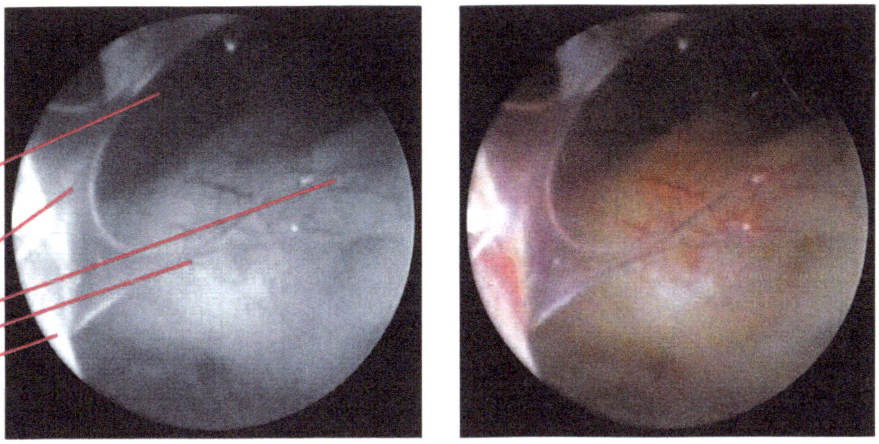

Fig. 81 Scope rotated clockwise to look straight ahead at the medial wall of the tympanic cavity (rigid scope, 1.9 mm, 30°).

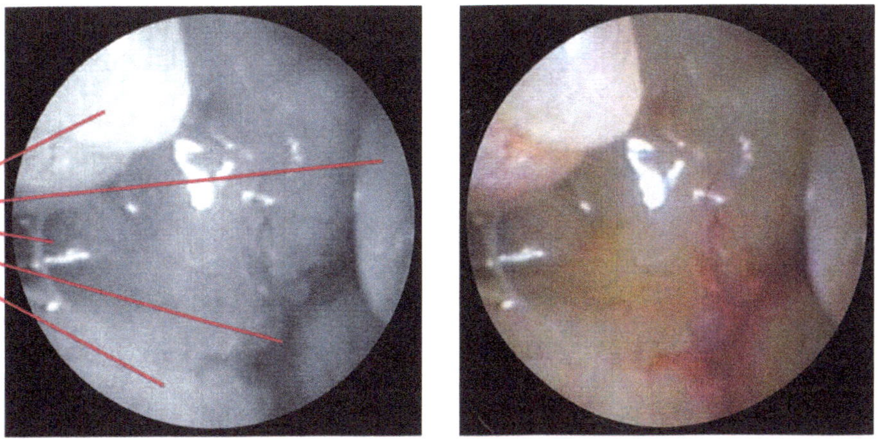

B 2.4 Looking up and forward: Anterior epitympanic recess

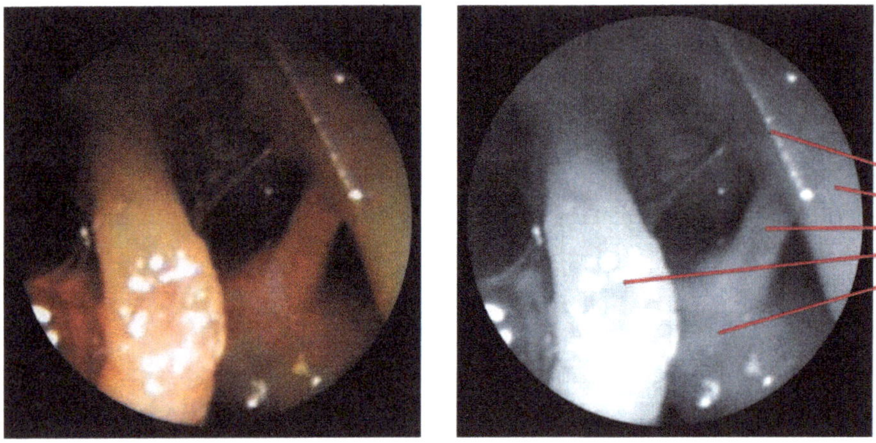

Fig. 82 Overview of the anterior epitympanic recess (rigid scope, 1.9 mm, 60°, right ear).

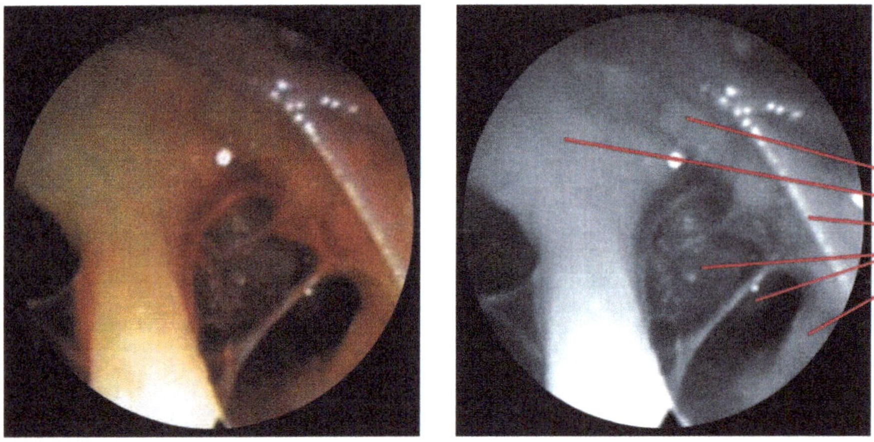

Fig. 83 Anterior epitympanic recess seen through the anterior tympanic isthmus (rigid scope, 1.9 mm, 60°).

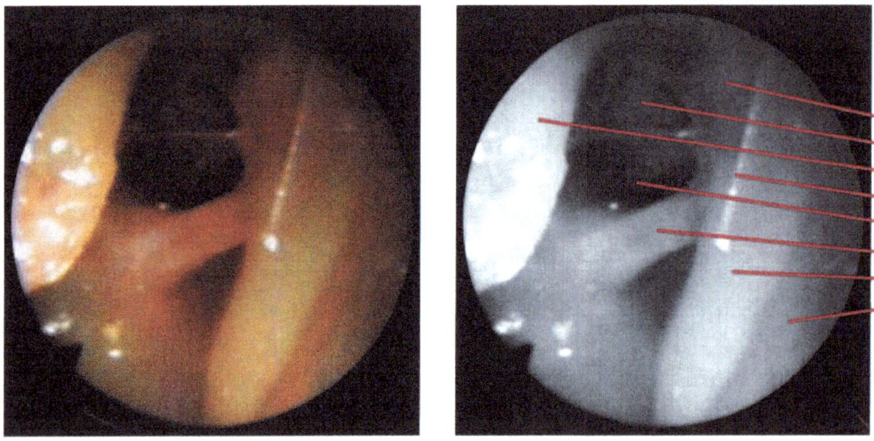

Fig. 84 Scope looks at the insertion of the tendon of the tensor tympani (note "figure of 7"), (rigid scope, 1.9 mm, 60°).

B 2.5 Looking up and backward: Aditus ad antrum

Fig. 85 Overview in a superior direction (rigid scope, 1.9 mm, 60°, right ear).

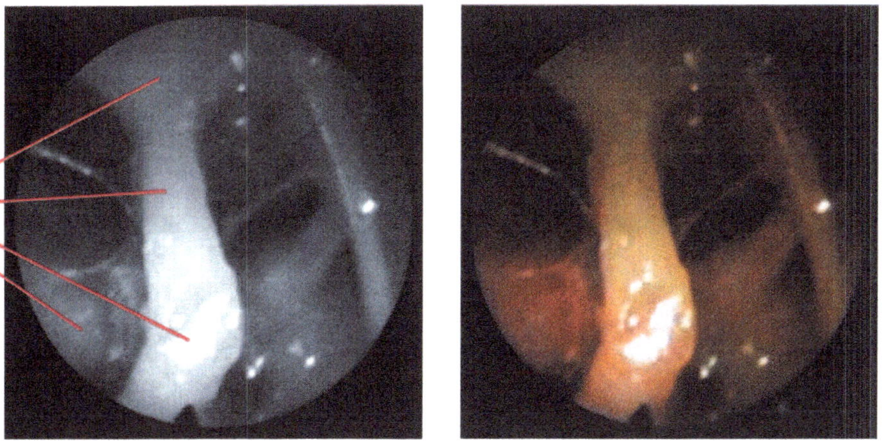

Fig. 86 View into the incudal fossa (rigid scope, 1.9 mm, 60°).

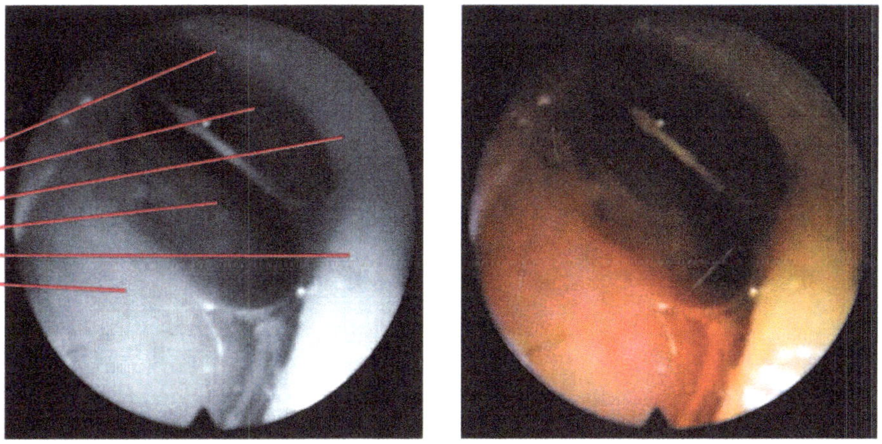

Fig. 87 View into the aditus ad antrum (rigid scope, 1.9 mm, 60°).

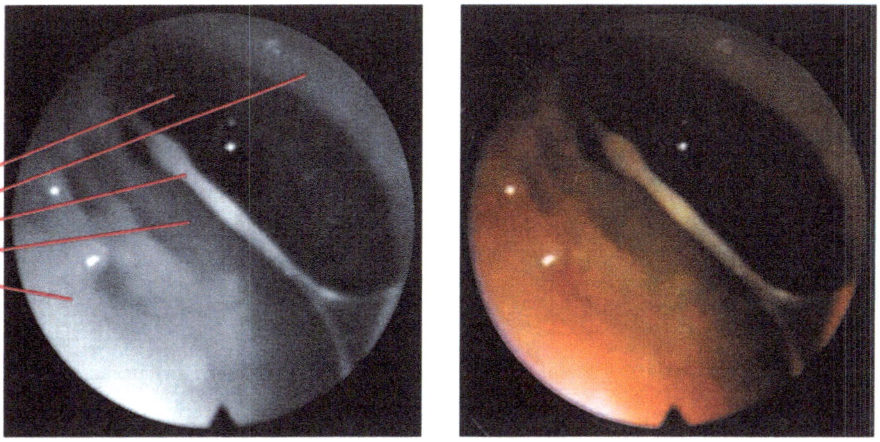

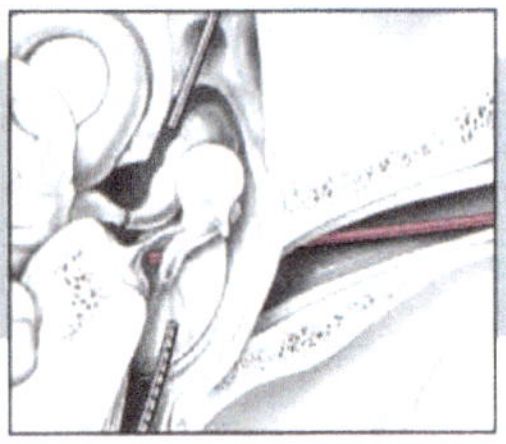

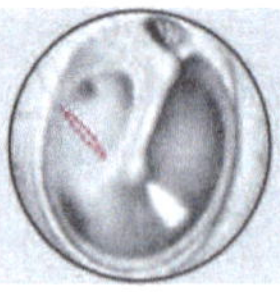

B 2.6 Looking forward and down: Protympanum and hypotympanum

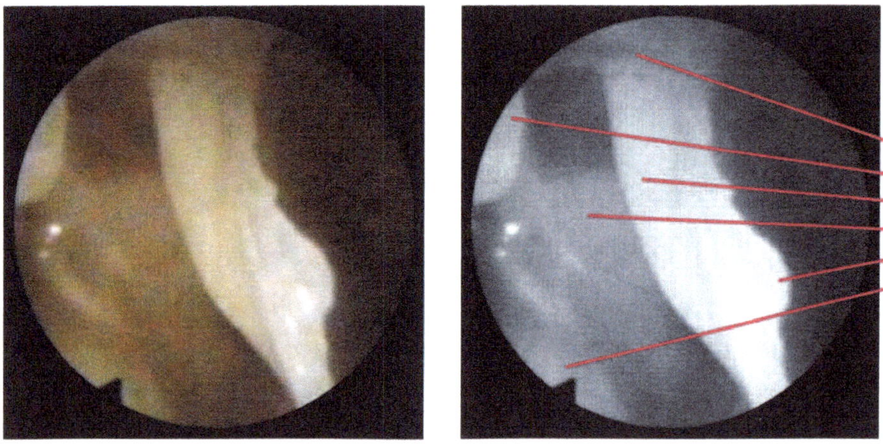

Fig. 88 View in an anterior direction (rigid scope, 1.9 mm, 60°, right ear).

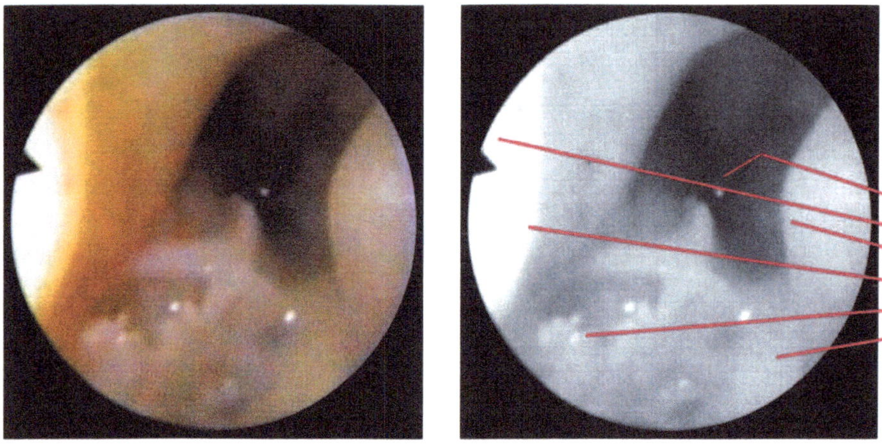

Fig. 89 Exploration of the anterior inferior tympanic cavity with the scope rotated clockwise to look toward the EUSTACHIAN tube (rigid scope, 1.9 mm, 60°).

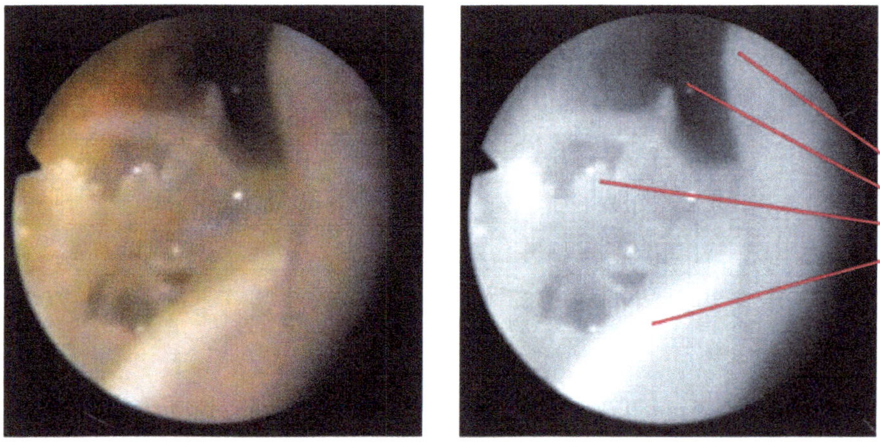

Fig. 90 Scope slightly tipped downward to look at the anterior hyptympanum (rigid scope, 1.9 mm, 60°).

B 3 Posterior inferior transtympanic approach

1.9 mm rigid scope

B 3.1	Site of approach
B 3.2.1	Approaching the round window niche (30° scope)
B 3.2.2	Secondary tympanic membrane (30° scope)
B 3.3.1	Stapes (30° scope)
B 3.3.2	View into vestibule (stapes removed) (30° scope)
B 3.4	Posterior wall of tympanic cavity (30° scope)
B 3.5	Posterior wall (60° scope)
B 3.6	Stapes seen from below (60° scope)
B 3.7	Roof and medial wall of protympanum (60° scope)
B 3.8	Floor of protympanum and hypotympanum (60° scope)

B 3.1 Site of approach

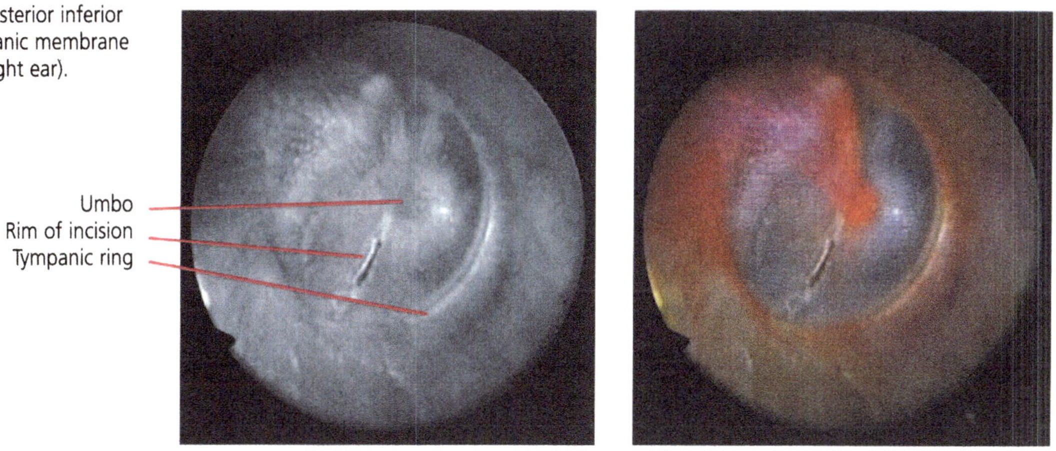

Fig. 91 Incision in the posterior inferior quadrant of a right tympanic membrane (rigid scope, 4 mm, 0°, right ear).

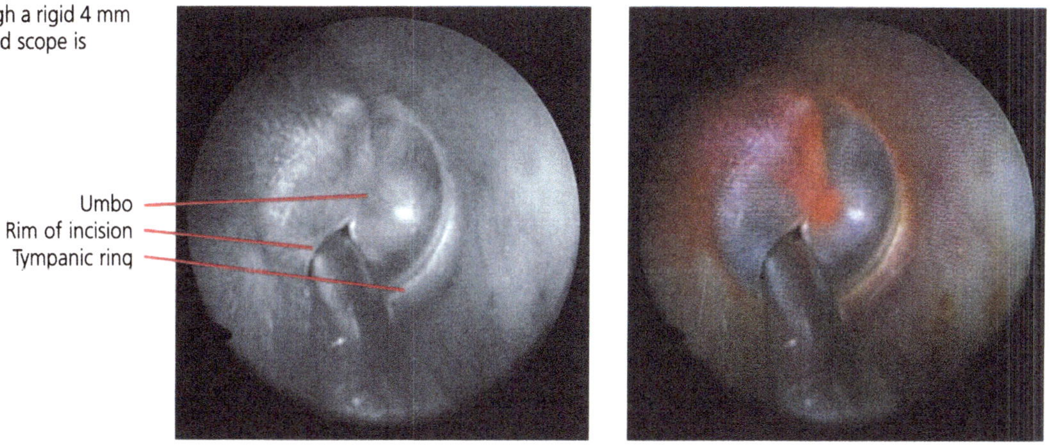

Fig. 92 Monitored through a rigid 4 mm 0° scope, the 1.9 mm rigid scope is introduced.

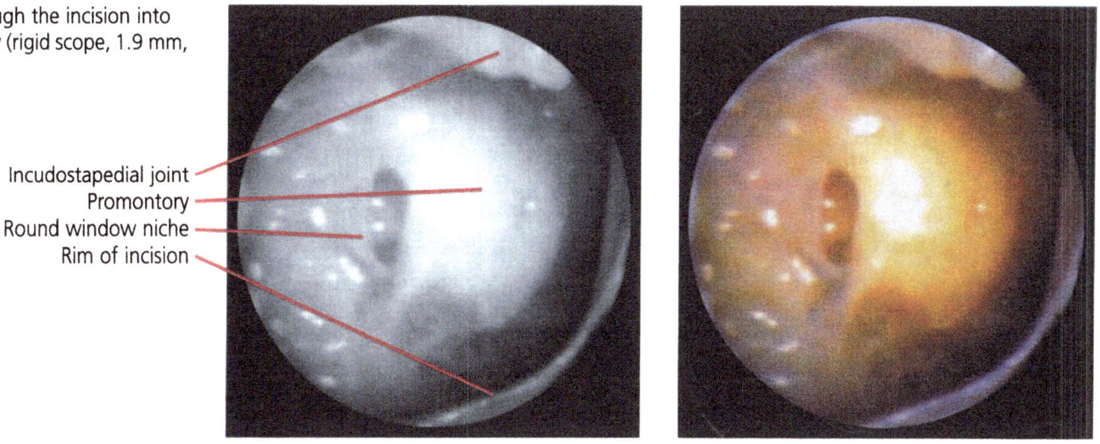

Fig. 93 View through the incision into the tympanic cavity (rigid scope, 1.9 mm, 0°, wide angle).

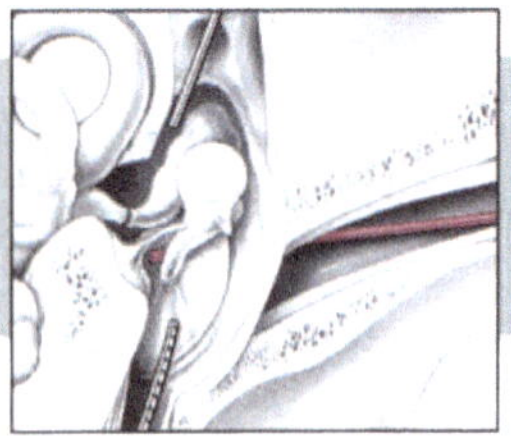

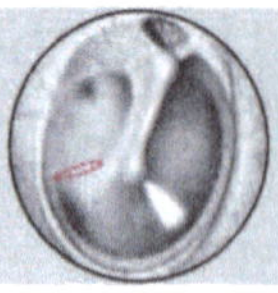

B 3.2.1 Approaching the round window niche

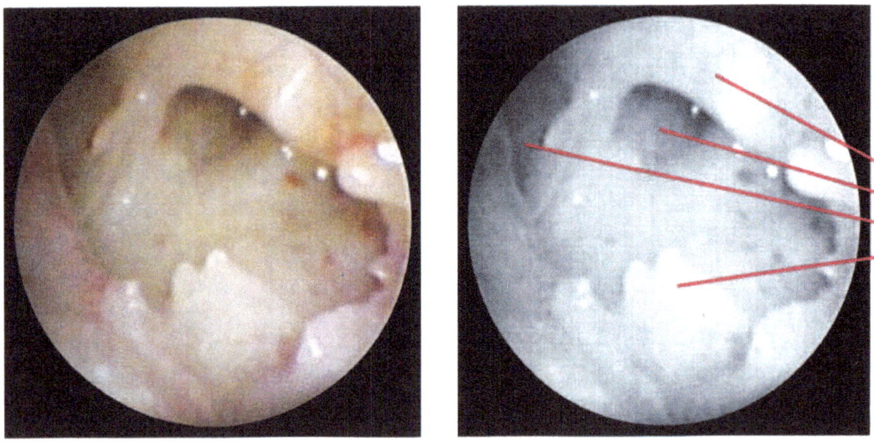

Fig. 94 Overview of round window niche and surrounding structures in an inferior posterior direction (rigid scope, 1.9 mm, 30°, right ear).

Promontory
Round window niche
Tympanic sinus
Trabecula of hypotympanum

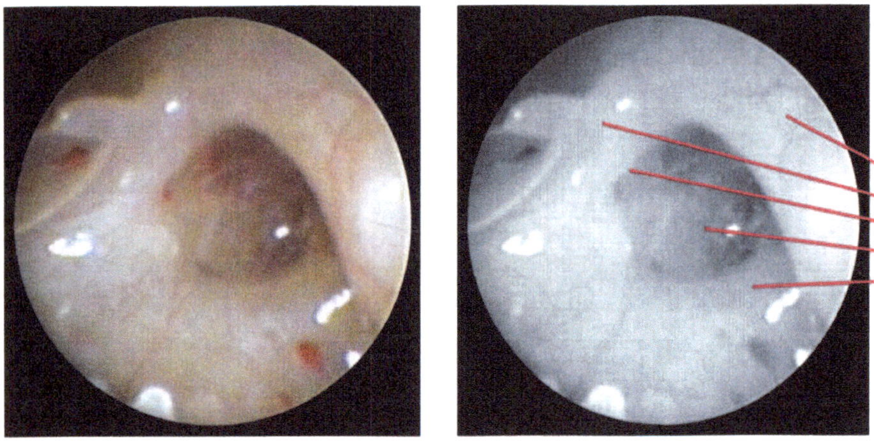

Fig. 95 Structures bordering the round window niche (rigid scope, 1.9 mm, 30°).

Promontory
Subiculum
Postis posterior
Round window niche
Postis anterior

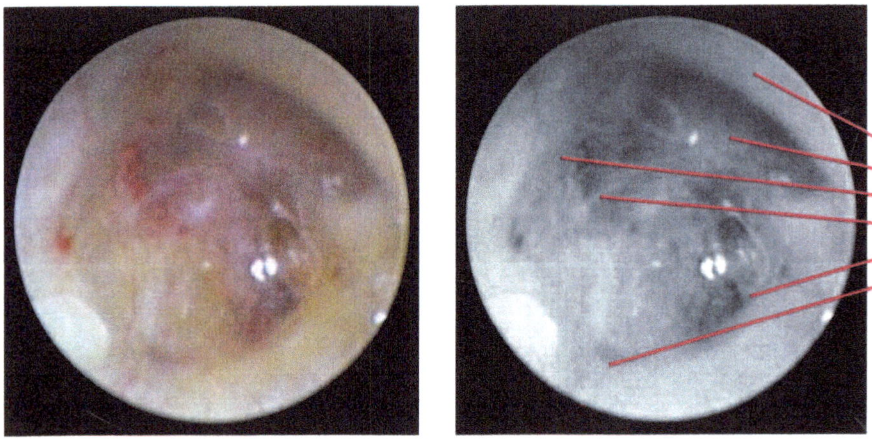

Fig. 96 Scope closer to secondary tympanic membrane (rigid scope, 1.9 mm, 30°).

Promontory
Secondary tympanic membrane
Postis posterior
Mucosal fold
Postis anterior
Floor of round window niche

B 3.2.2 Step-by-step entry through secondary tympanic membrane

Images taken through transtympanic incision after reflection of tympanomeatal flap

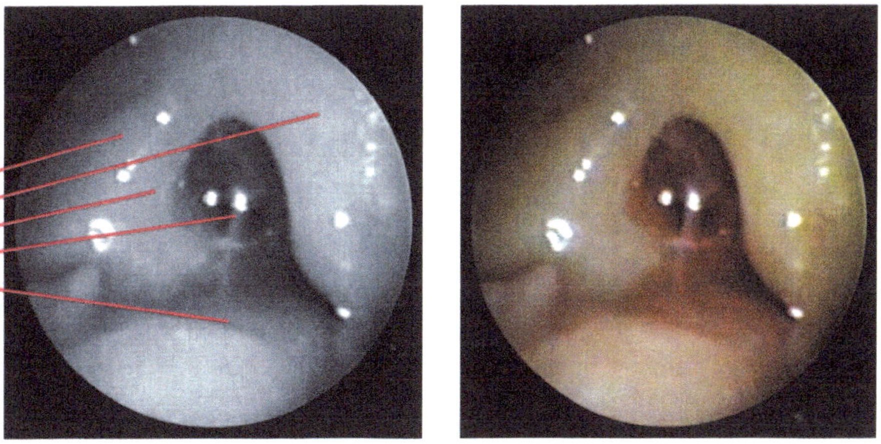

Fig. 97 Boundaries of round window niche (rigid scope, 1.9 mm, 30°).

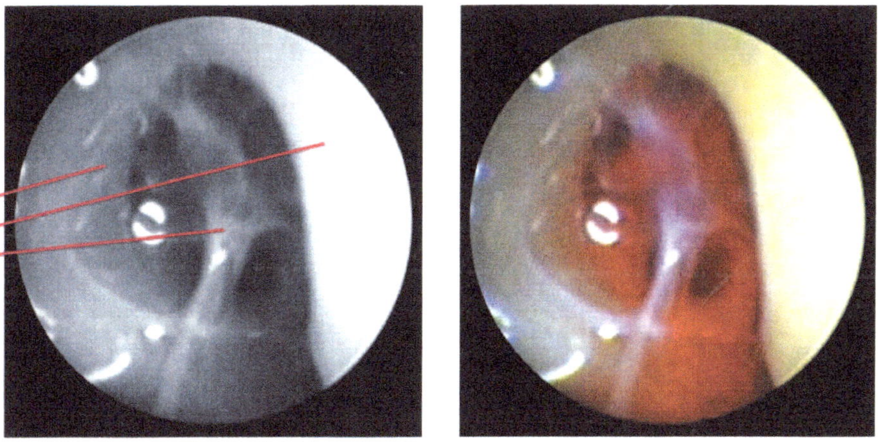

Fig. 98 Round window niche covered by mucosal fold (rigid scope, 1.9 mm, 30°).

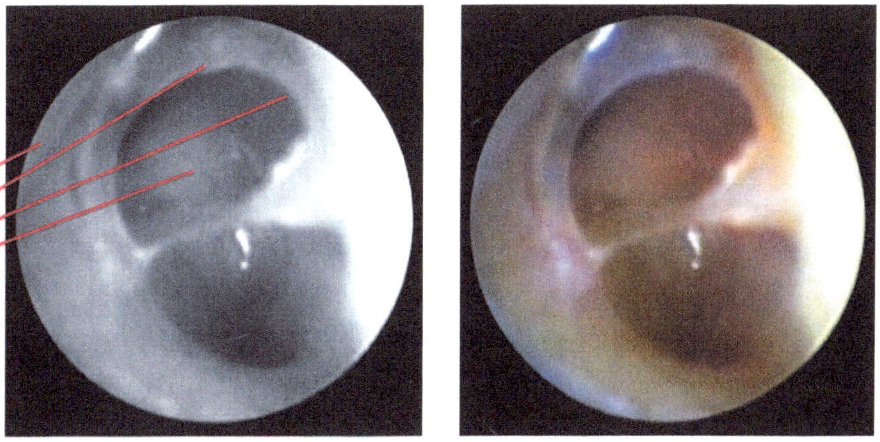

Fig. 99 Perforation of the mucosal fold (rigid scope, 1.9 mm, 30°).

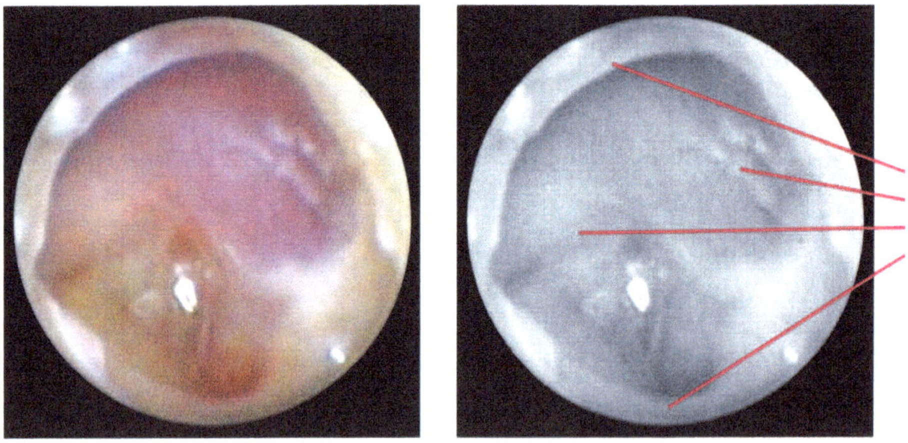

Fig. 100 View of the secondary tympanic membrane with the mucosal fold removed (rigid scope, 1.9 mm, 30°, right ear).

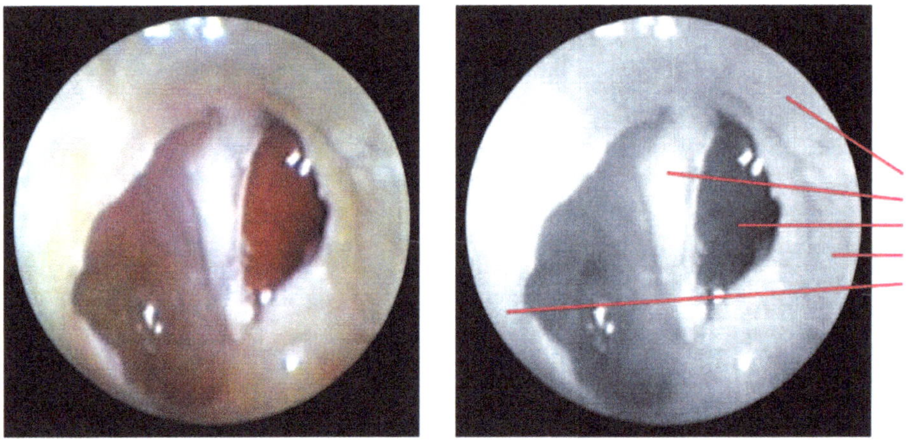

Fig. 101 Secondary tympanic membrane partly opened (rigid scope, 1.9 mm, 30°).

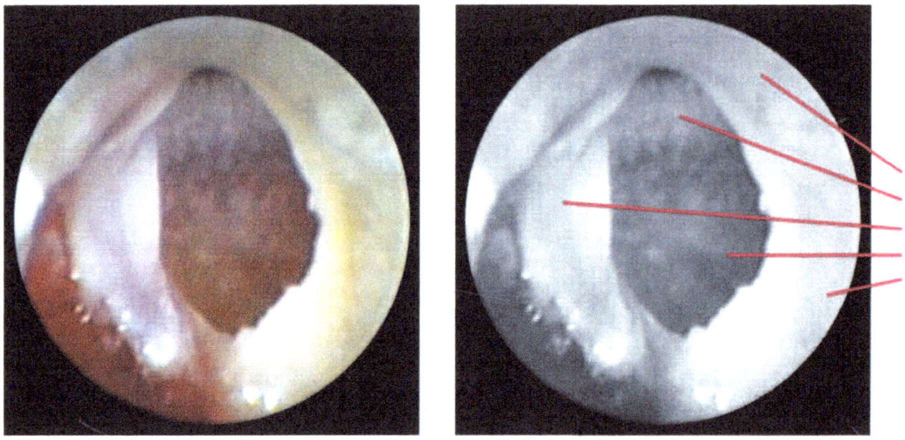

Fig. 102 Round window wide open (rigid scope, 1.9 mm, 30°).

B 3.3.1 Stapes

Fig. 103 Overview with the incudostapedial joint in the center (rigid scope, 1.9 mm, 30°, right ear).

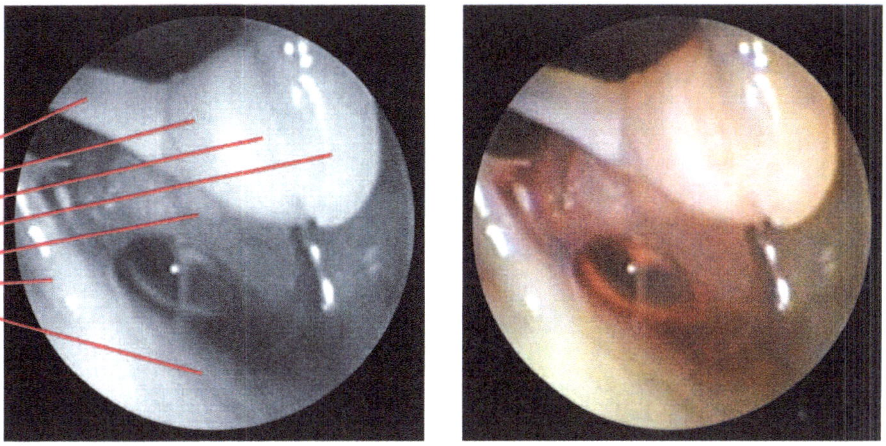

Fig. 104 Closer view of the body of stapes (rigid scope, 1.9 mm, 30°).

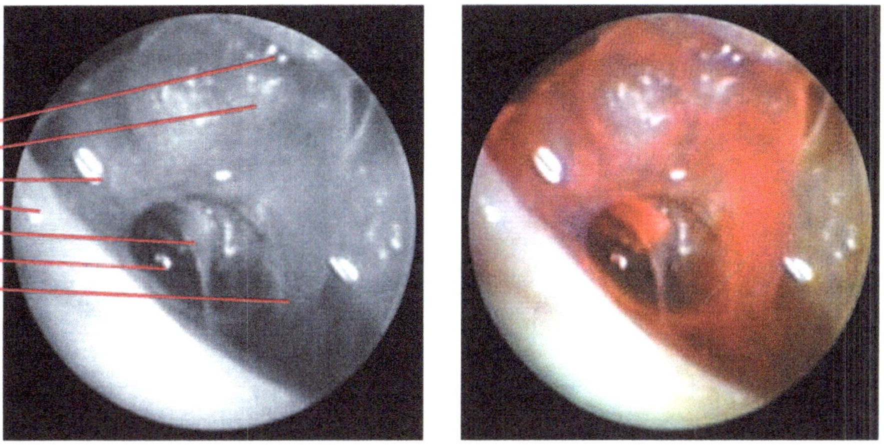

Fig. 105 Scope rotated clockwise to see the anterior limb of stapes (rigid scope, 1.9 mm, 30°).

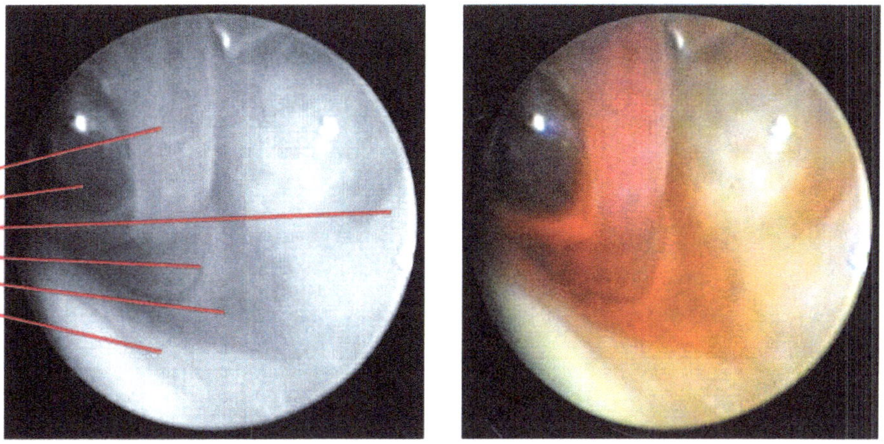

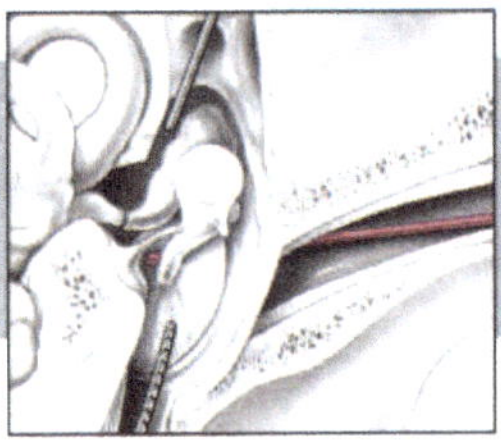

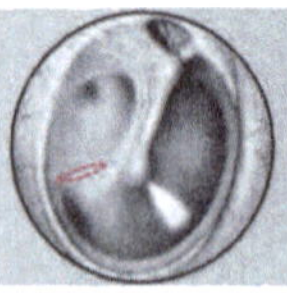

B 3.3.2 View into vestibule (stapes removed)

Stapes removed along transmeatal approach, images taken through transtympanic incision after reflection of the tympanomeatal flap

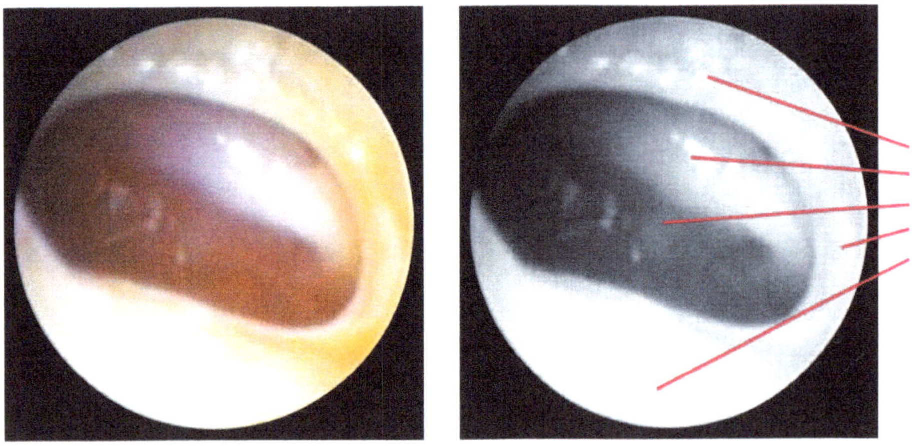

Fig. 106 Overview of the oval window (rigid scope, 1.9 mm, 30°, right ear).

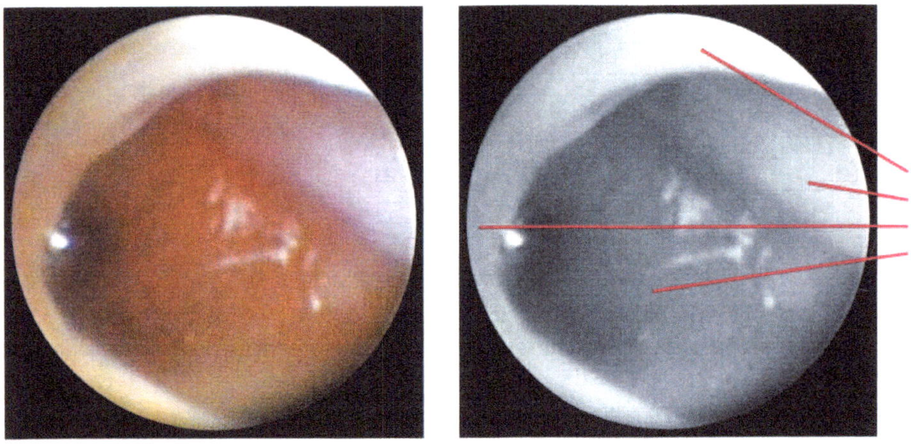

Fig. 107 Counterclockwise rotation of the scope brings into view the posterior half of the oval window (rigid scope, 1.9 mm, 30°).

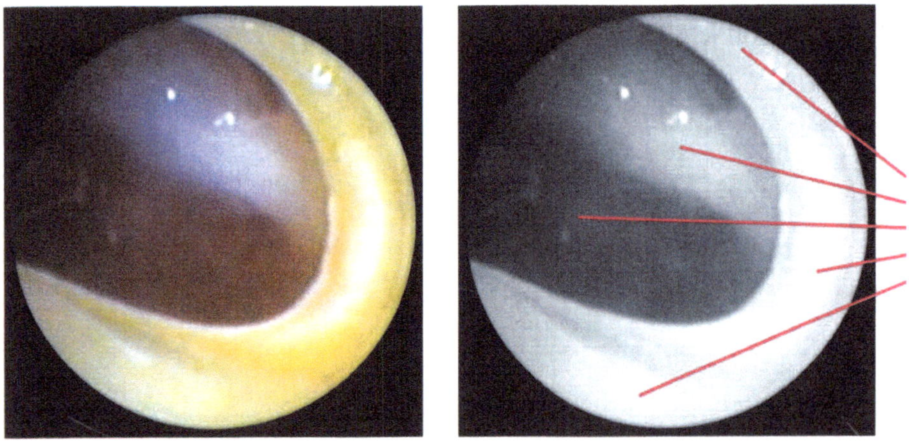

Fig. 108 Clockwise rotation of the scope brings into view the anterior half of the oval window (compare Fig. 107), (rigid scope, 1.9 mm, 30°).

B 3.4 Posterior wall of tympanic cavity

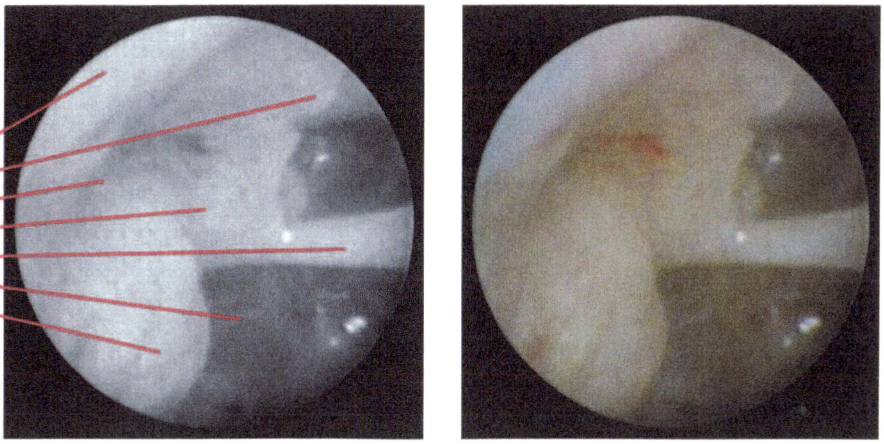

Fig. 109 Overview of the posterior wall with the lateral tympanic recess (rigid scope, 1.9 mm, 30°, right ear).

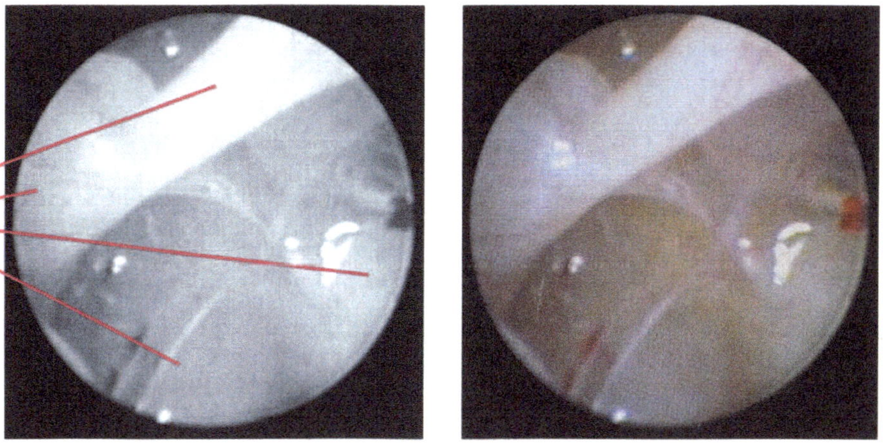

Fig. 110 Slight counterclockwise rotation of the scope provides an overview of the tympanic sinus (rigid scope, 1.9 mm, 30°).

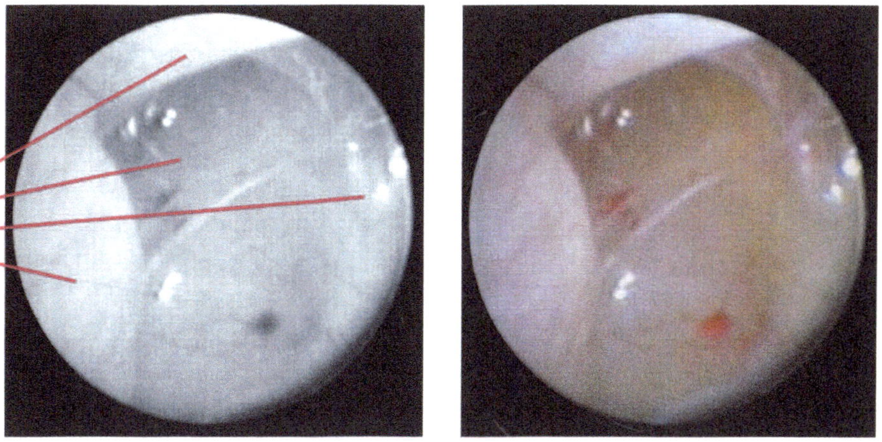

Fig. 111 With the scope directed downward from the stapedius tendon the entire tympanic sinus can be seen (rigid scope, 1.9 mm, 30°).

B 3.5 Looking back: Posterior wall

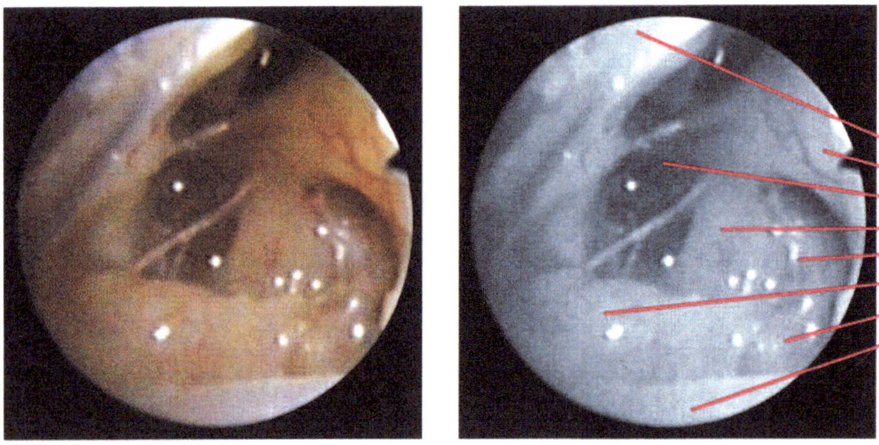

Fig. 112 Overview of the posterior part of the tympanic cavity (rigid scope, 1.9 mm, 60°, right ear).

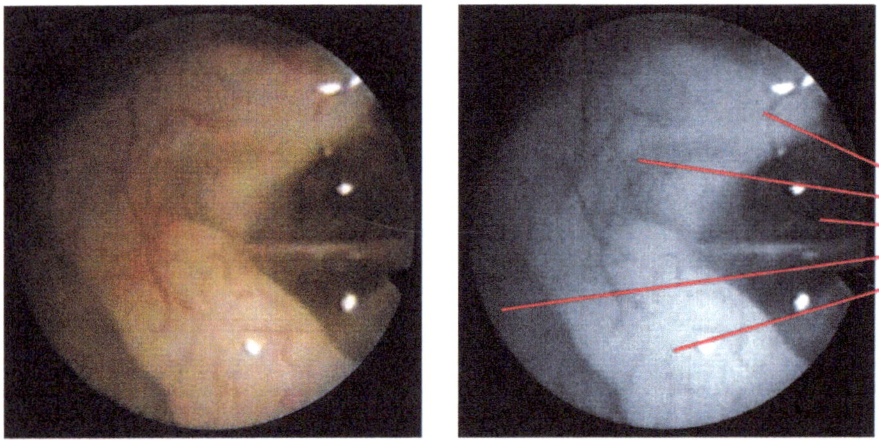

Fig. 113 View of the lateral tympanic recess with the scope in the plane of the incision (rigid scope, 1.9 mm, 60°).

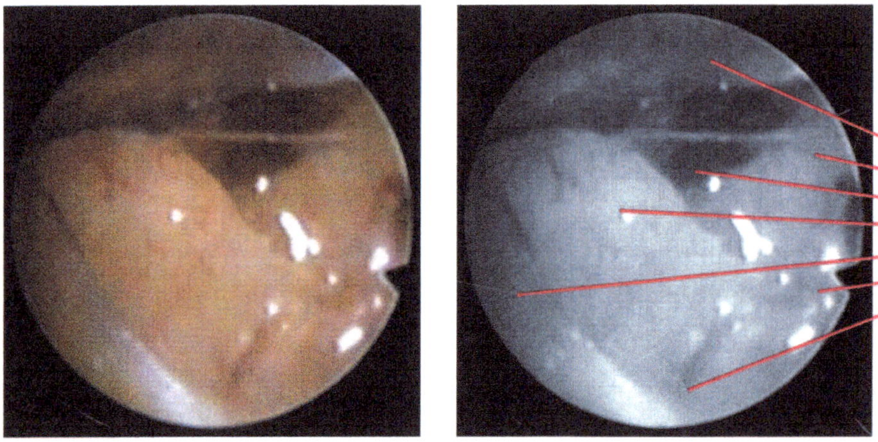

Fig. 114 Advancing the scope forward shows the posterior hypotympanum (rigid scope, 1.9 mm, 60°).

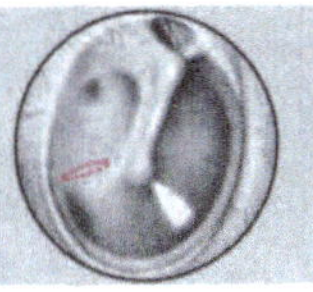

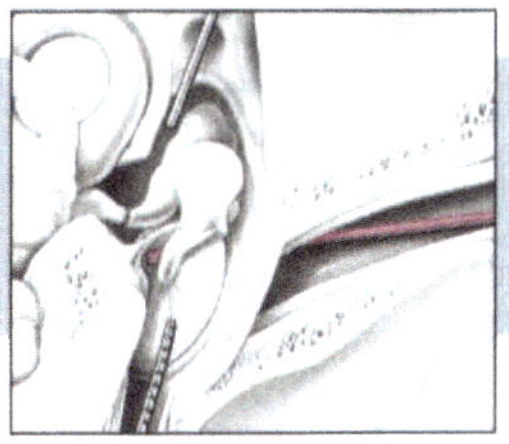

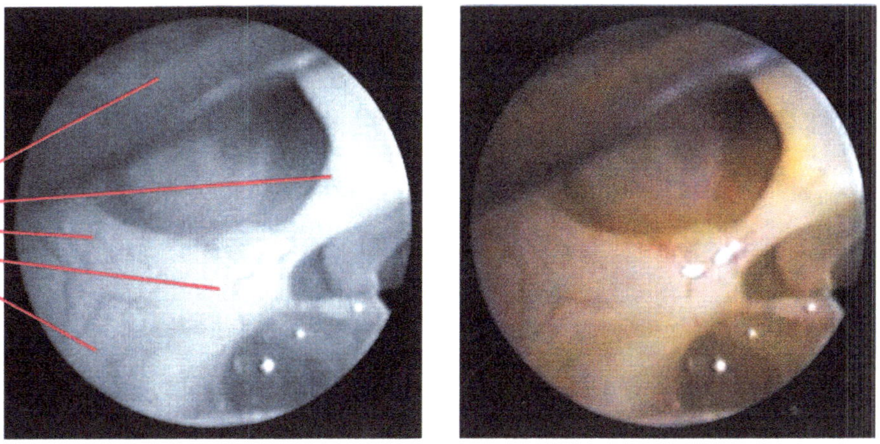

Fig. 115 The pyramidal eminence and the surrounding structures (rigid scope, 1.9 mm, 60°, right ear).

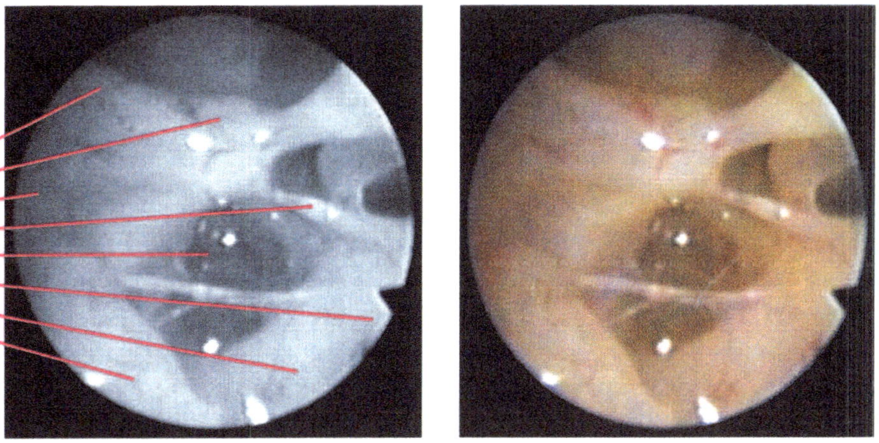

Fig. 116 Boundaries of the tympanic sinus (rigid scope, 1.9 mm, 60°).

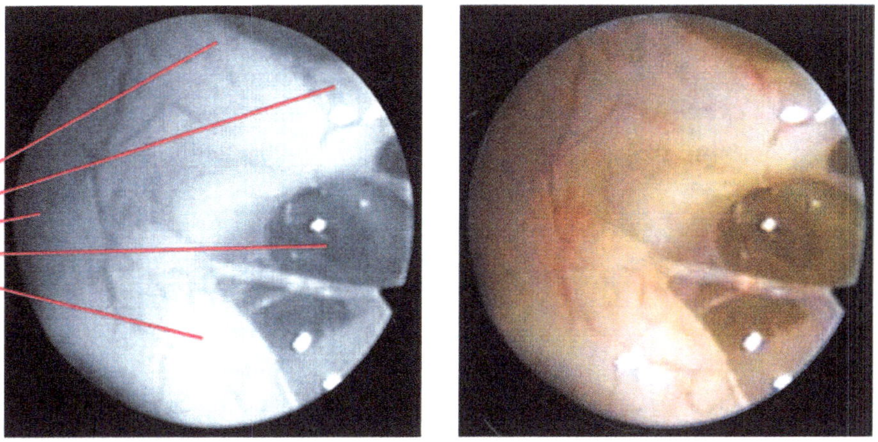

Fig. 117 Inferior part of the posterior wall of the tympanic cavity (rigid scope, 1.9 mm, 60°).

B 3.6 Looking up: Stapes seen from below

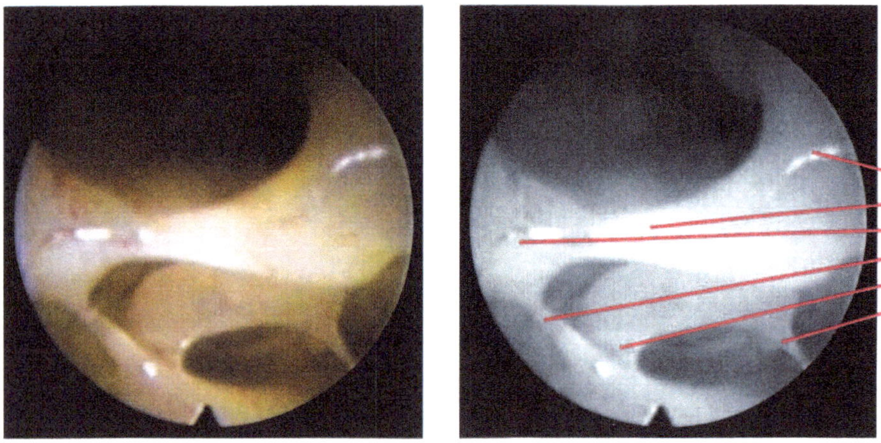

Fig. 118 Overview with the insertion of the tendon of stapedius (rigid scope, 1.9 mm, 60°, right ear).

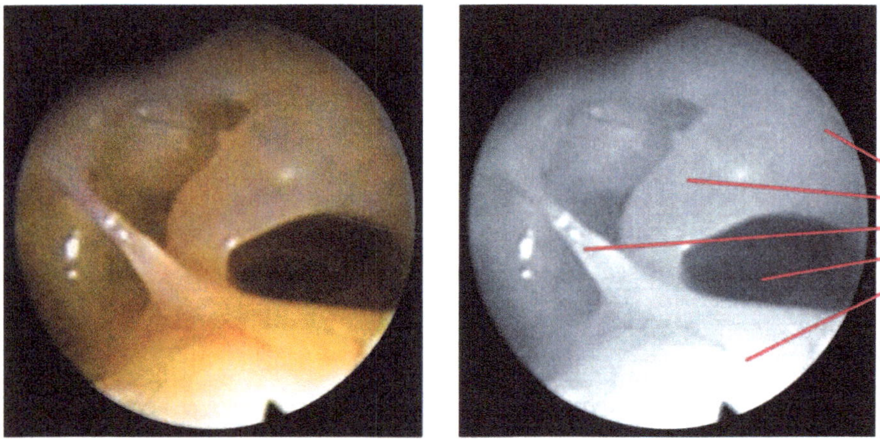

Fig. 119 Turning the scope counter-clockwise and looking posteriorly the posterior limb of stapes is seen (rigid scope, 1.9 mm, 60°).

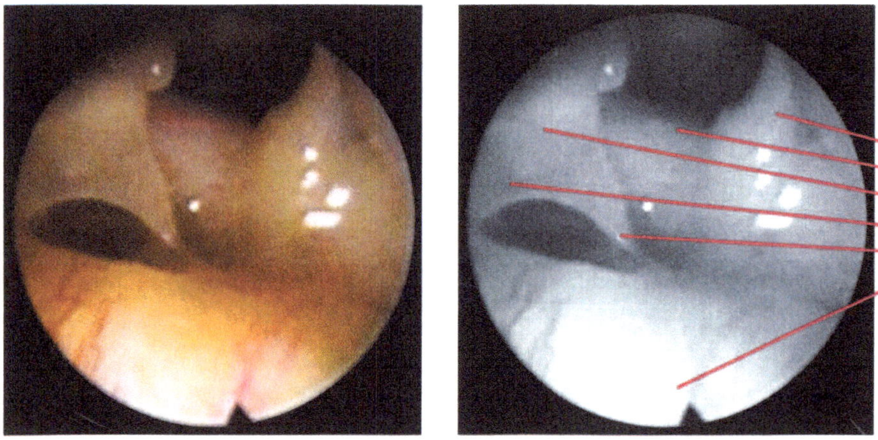

Fig. 120 Same rotation as in Fig. 119, but with the scope tipped anteriorly shows the anterior limb of stapes (rigid scope, 1.9 mm, 60°).

B 3.7 Looking forward: Roof and medial wall of protympanum

Fig. 121 Scope looking at the tendon of the tensor tympani (rigid scope, 1.9 mm, 60°, right ear).

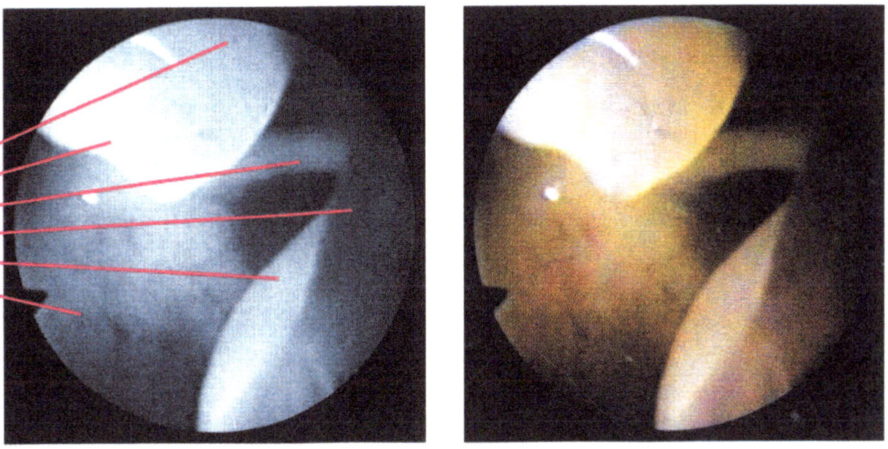

Fig. 122 Inner surface of the tympanic membrane; umbo is also seen (rigid scope, 1.9 mm, 60°).

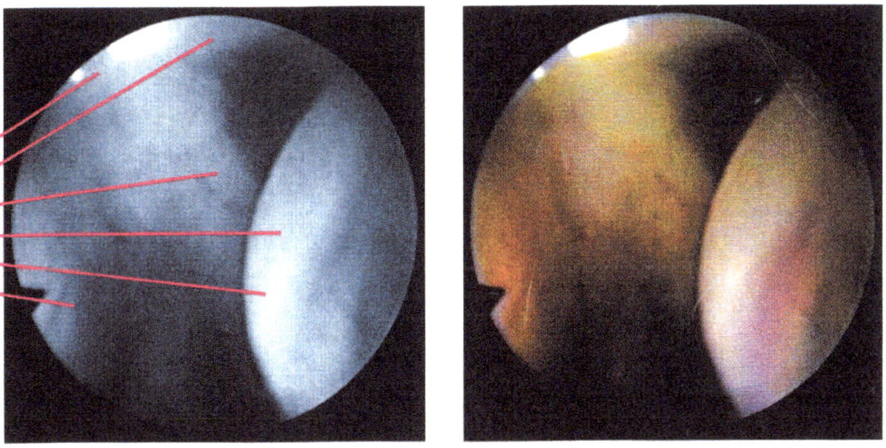

Fig. 123 View into the protympanum close to EUSTACHIAN tube opening (rigid scope, 1.9 mm, 60°).

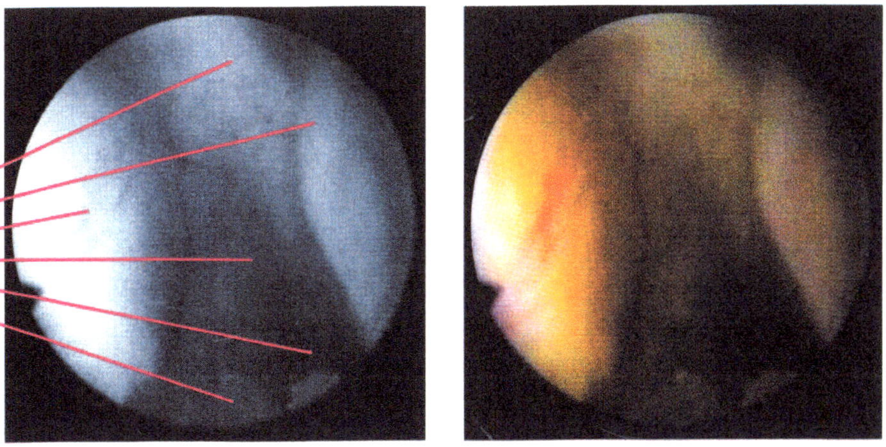

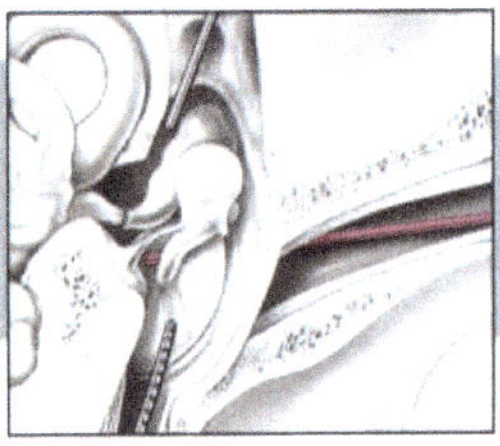

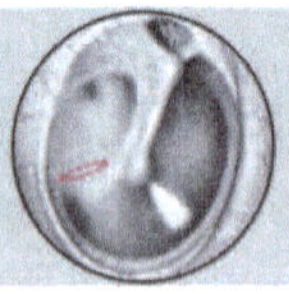

B 3.8 Looking down: Floor of protympanum and hypotympanum

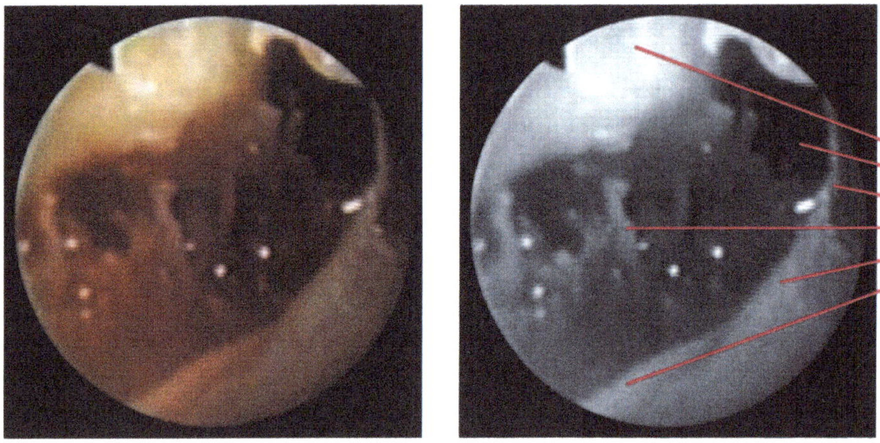

Fig. 124 Bottom of tympanic cavity and inner surface of tympanic membrane (rigid scope, 1.9 mm, 60°, right ear).

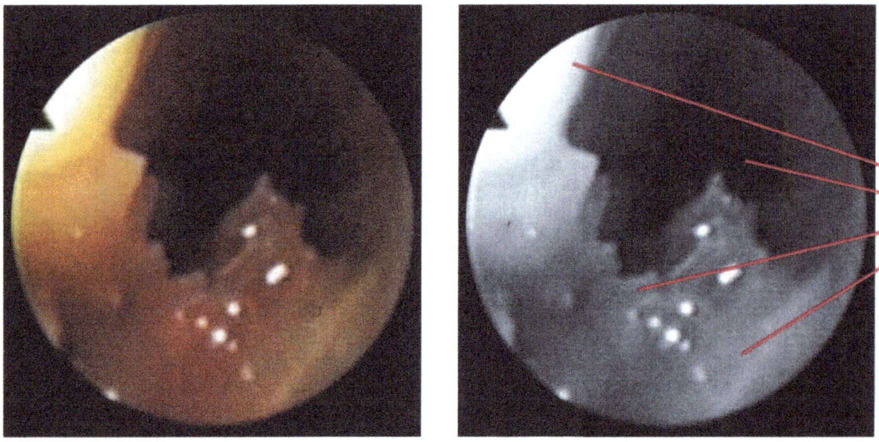

Fig. 125 Bottom of protympanum (rigid scope, 1.9 mm, 60°).

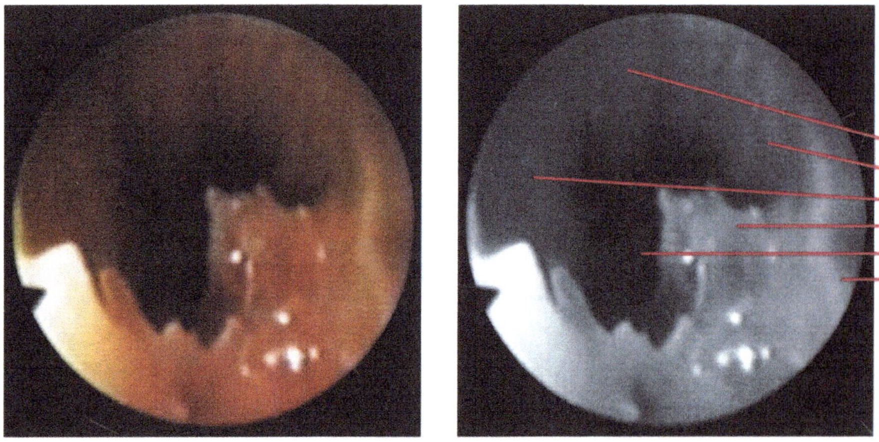

Fig. 126 Tympanic opening of EUSTACHIAN tube (rigid scope, 1.9 mm, 60°).

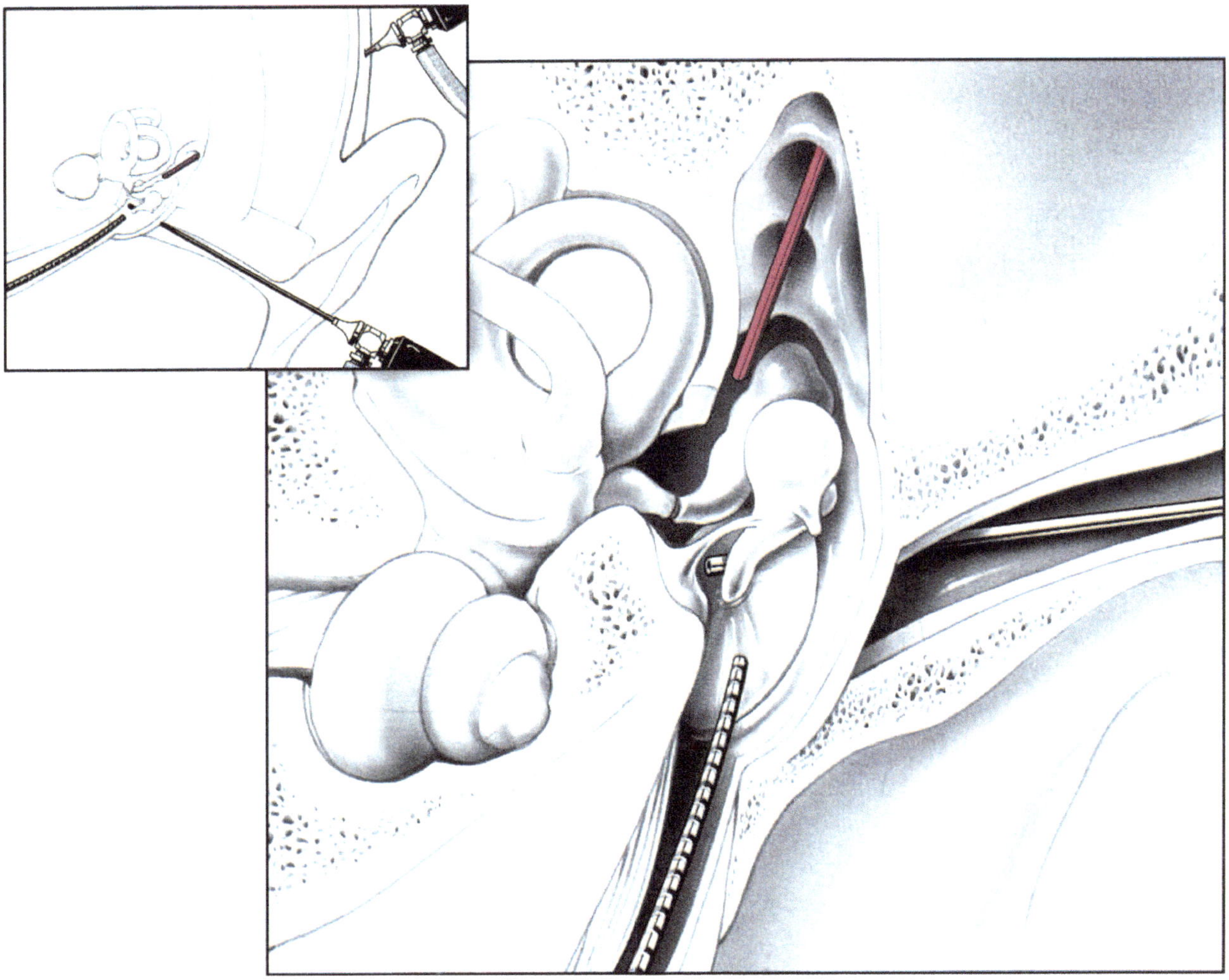

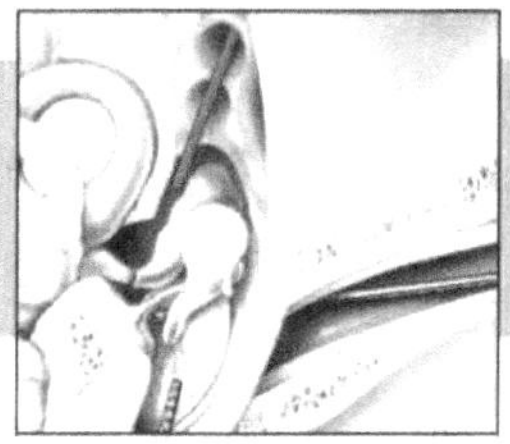

Transmastoid approach to the middle ear

From the mastoid air cells the middle ear is best approached through the antrum mastoideum. Partial mastoidectomy (preserving the posterior wall of the auditory canal) leaves 2 potential approaches to the middle ear from posterior: (1) through the antrum mastoideum into the epitympanum and (2) through the facial recess into the mesotympanum. Along these posterior approaches the ossicular chain and the inner surface of the tympanic membrane can be explored. The oval window and the footplate are not seen, because they lie deep to the facial canal.

(1) The narrow aditus ad antrum can only be negotiated by a flexible steerable scope with a maximum outer diameter of 1.2 mm. The structures surrounding it, i.e. the lateral semicircular canal, the short limb of the incus and the facial canal, form a triangle with the apex pointing downward as described by Anson and Donaldson. Through the flexible scope the cochleariform process and the ventrobasal tympanic opening of the Eustachian tube are seen. If this passage is too narrow, a 30° scope placed in the plane of the aditus will at least show the inferior structures such as the incudostapedial joint and the hypotympanum.

(2) Once the facial recess, which is bordered by the facial canal, the canal for the chorda tympani and the buttress, is opened by posterior tympanotomy, the mesotympanum with the incudostapedial joint and the tendon of the stapedius muscle are seen. The inner surface of the tympanic membrane comes into view laterally and the oval window niche medially. Looking down the secondary tympanic membrane may also be seen. Advancing the scope beneath the incudostapedial joint shows the hypotympanum, the protympanum and the tympanic opening of the Eustachian tube.

These approaches have obvious importance for postoperative second look studies particularly in children.

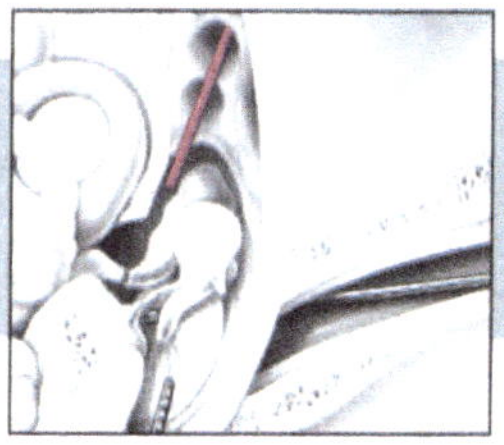

C 1 Transmastoid approach through antrum

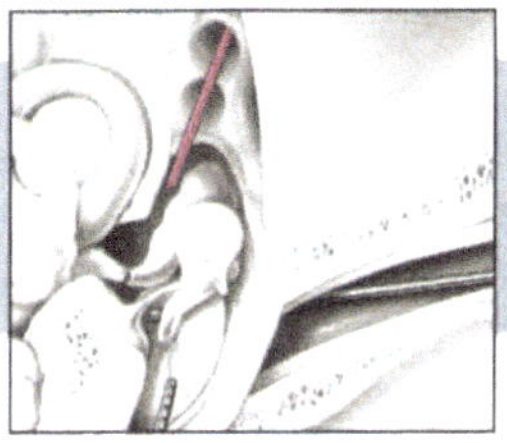

C 1.1 Site of approach

Fig. 127 Endoscopic overview gained through mastoidectomy and posterior tympanotomy (rigid scope, 4 mm, 0°, right ear).

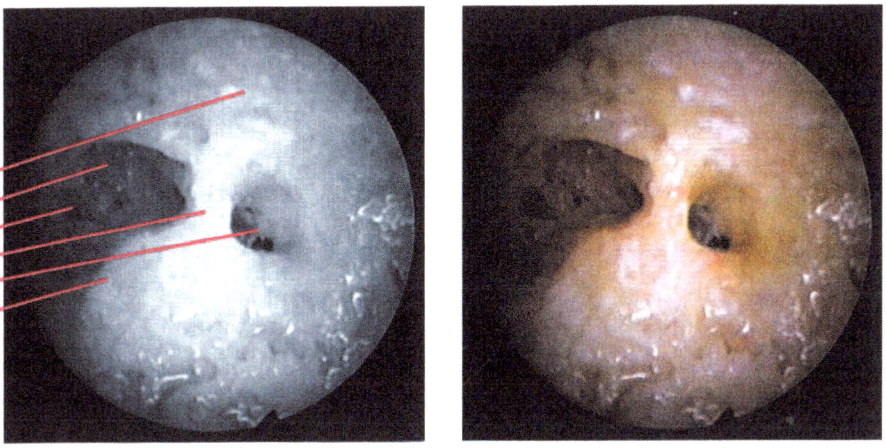

Fig. 128 Approach to the middle ear through antrum mastoideum (bordered by lateral semicircular canal, incus and buttress), (rigid scope, 4 mm, 0°).

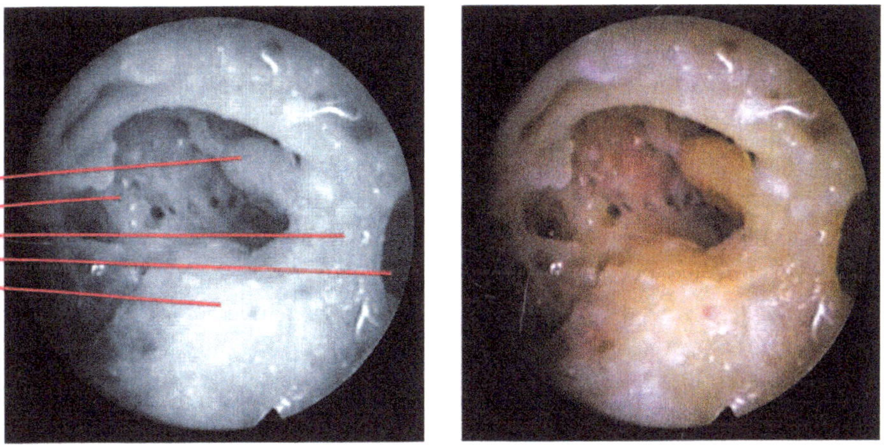

Fig. 129 Approach to the middle ear through facial recess (a triangle bordered by facial canal, canal for chorda tympani and buttress), (rigid scope, 4 mm, 0°).

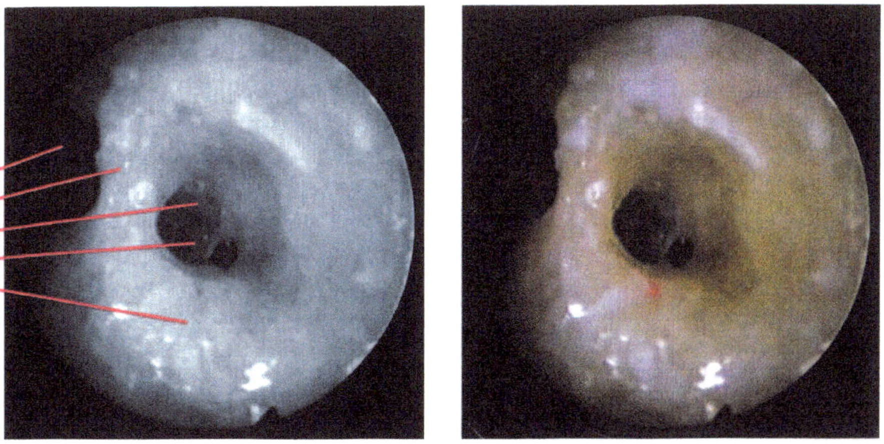

C 1.2 Antrum mastoideum

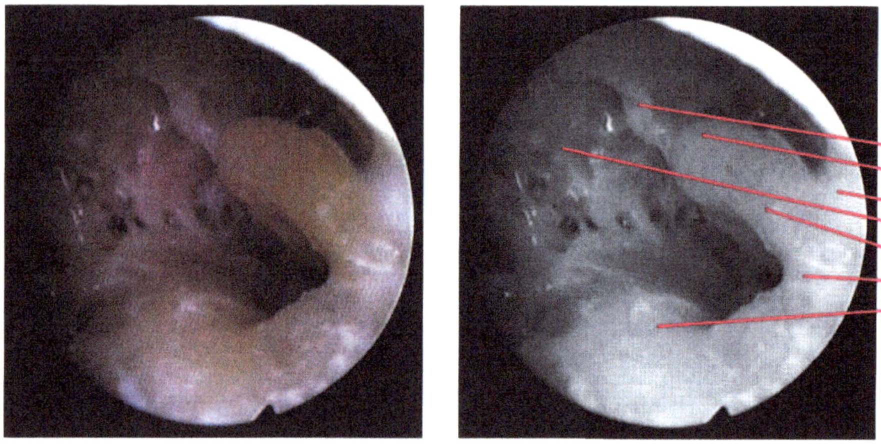

Fig. 130 Advancing a 0° scope straight ahead shows the incus in the center of the field of view (rigid scope, 4 mm, 0°, right ear).

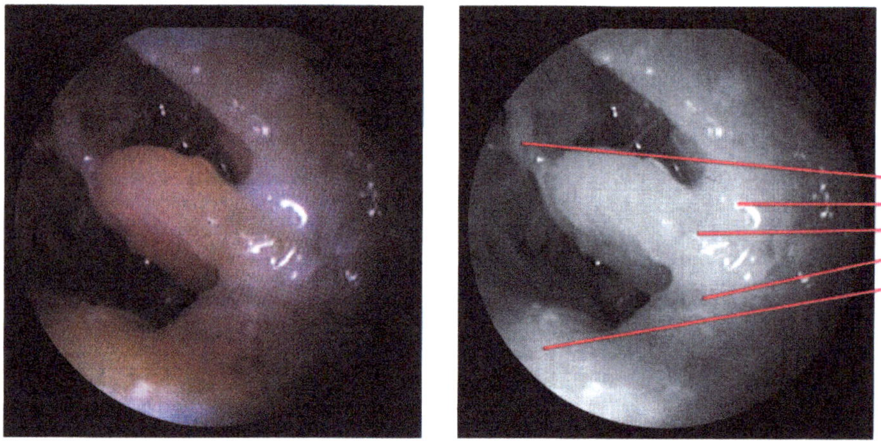

Fig. 131 30° scope advanced alongside tegmen tympani to look down at the short limb of the incus (rigid scope, 4 mm, 30°).

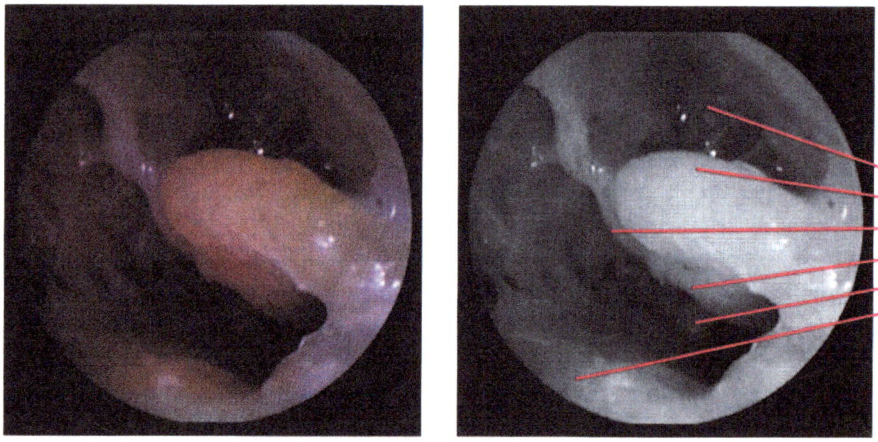

Fig. 132 With further progress of the scope the long limb of the incus and the cochleariform process are seen (rigid scope, 4 mm, 30°).

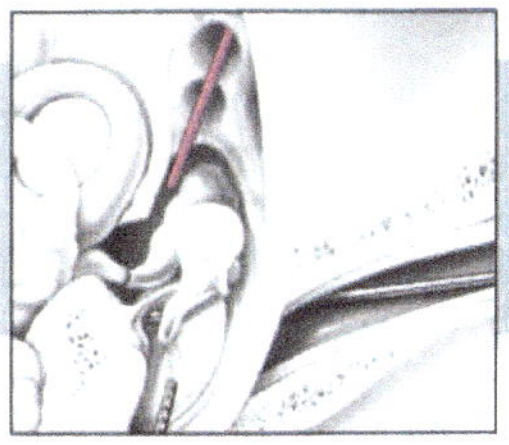

C 1.3 View into mesotympanum from the top down

Fig. 133 View into mesotympanum through the gap between the incus and the lateral semicircular canal (rigid scope, 2.7 mm, 30°, right ear).

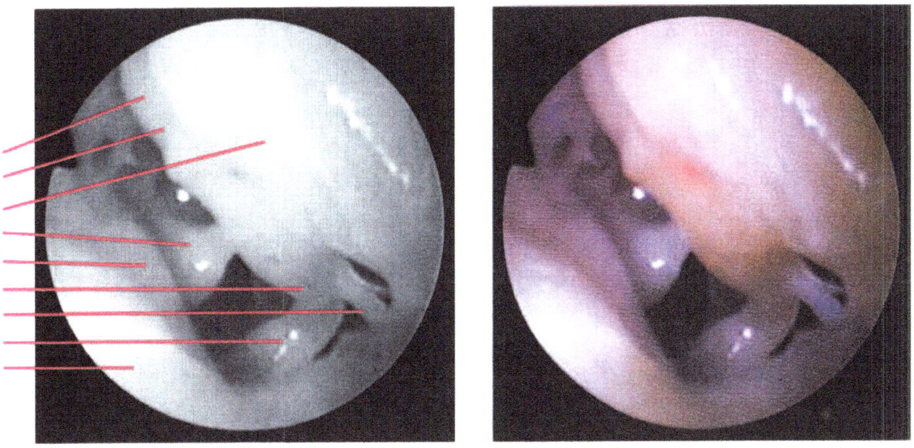

Fig. 134 Scope approaches the cochleariform process and the incudostapedial joint (rigid scope, 2.7 mm, 30°)

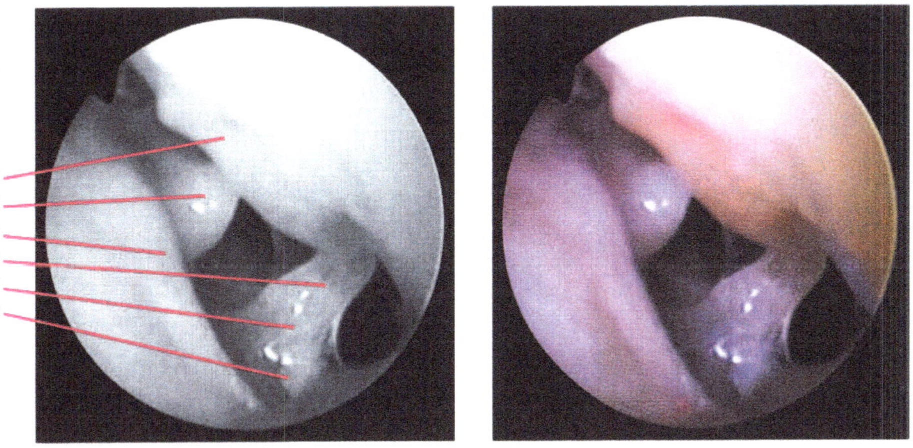

Fig. 135 Scope advanced to look into anterior epitympanic recess (rigid scope, 2.7 mm, 30°).

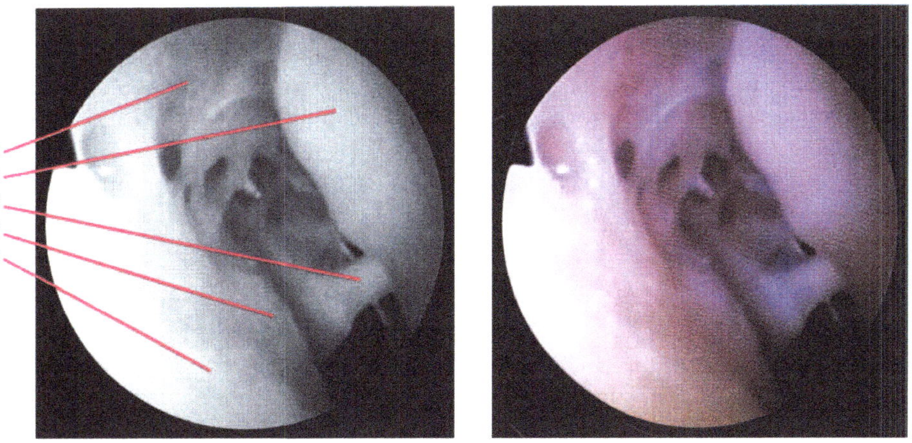

C 1.4.1 Incudostapedial joint and anterior tympanic isthmus

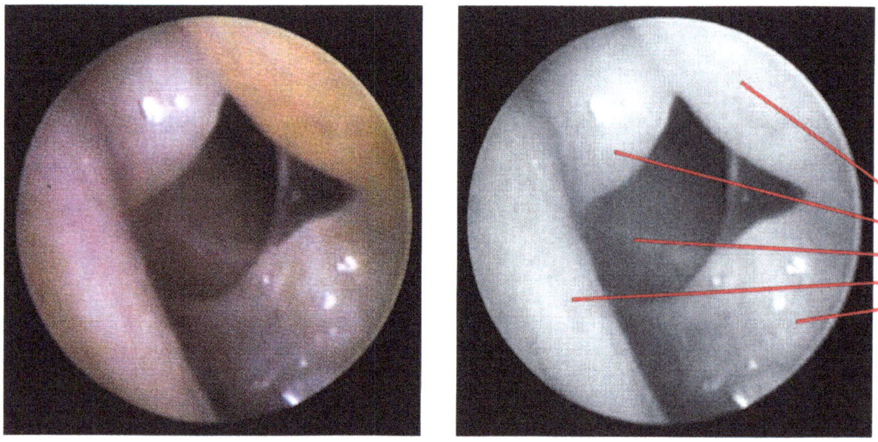

Fig. 136 View of promontory through typically rectangular opening (= anterior tympanic isthmus) between facial canal, cochleariform process, incus and joint (rigid scope, 1.9 mm, 30°, right ear).

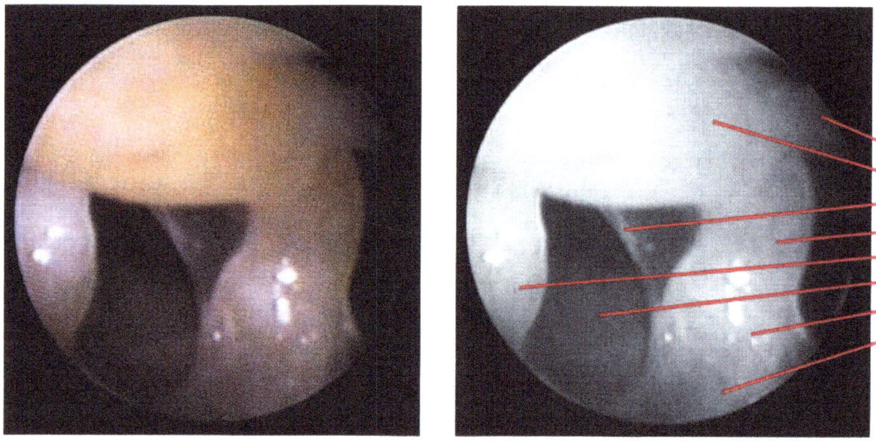

Fig. 137 Scope approaches anterior tympanic isthmus and focuses on incudostapedial joint (rigid scope, 1.9 mm, 30°).

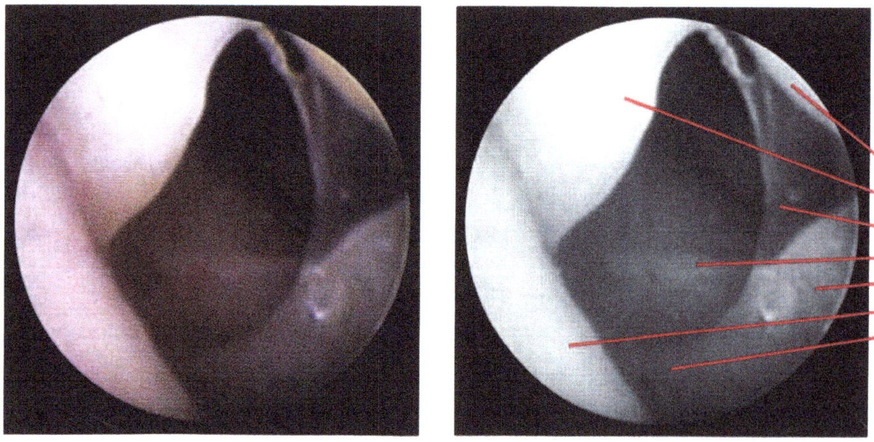

Fig. 138 Promontory and JACOBSON's nerve are seen in the center of the opening (rigid scope, 1.9 mm, 30°).

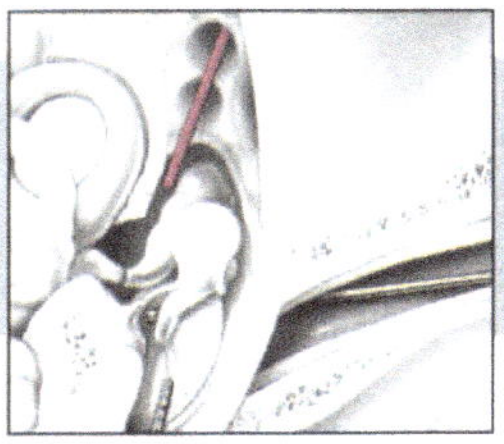

C 1.4.2 Cochleariform process and anterior epitympanic recess

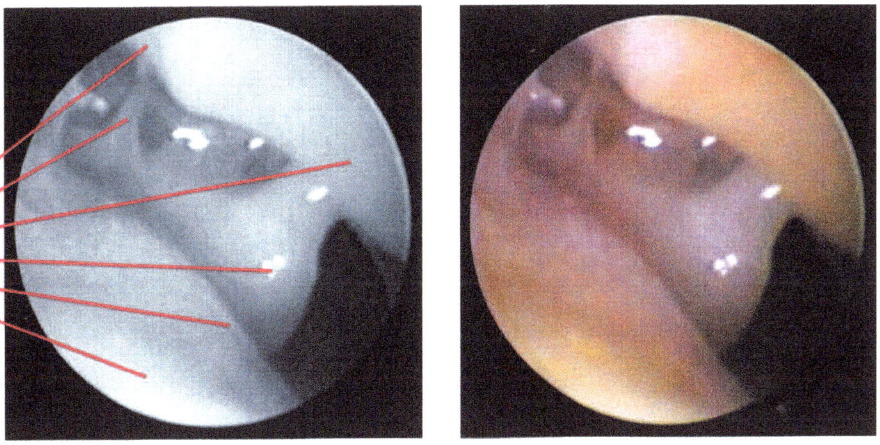

Fig. 139 View into anterior epitympanic recess (rigid scope, 1.9 mm, 30°, right ear).

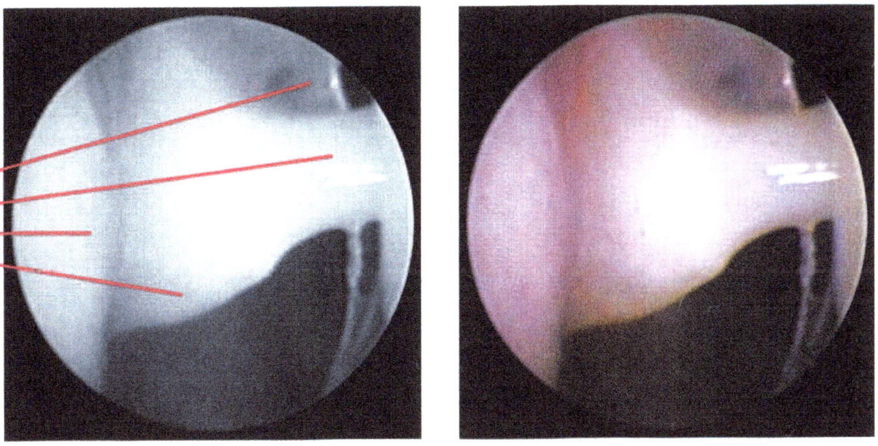

Fig. 140 Scope advanced to approach the cochleariform process (rigid scope, 1.9 mm, 30°).

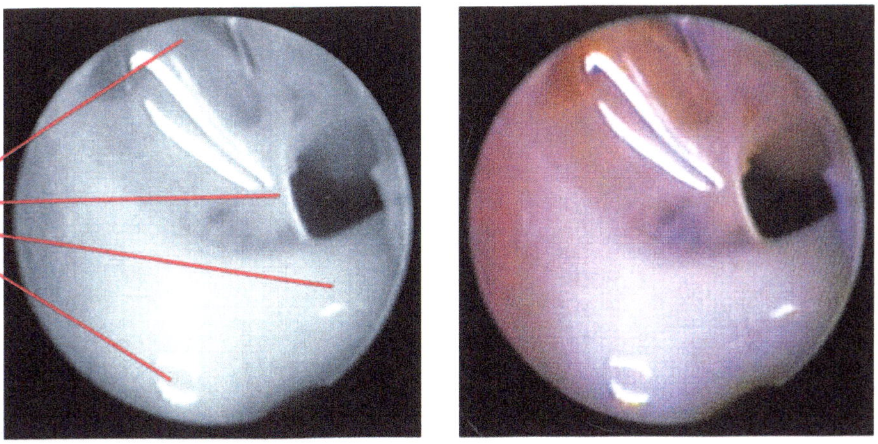

Fig. 141 View of an exceptional communication (=supratubal recess) between EUSTACHIAN tube and anterior epitympanic recess (rigid scope, 1.9 mm, 30°).

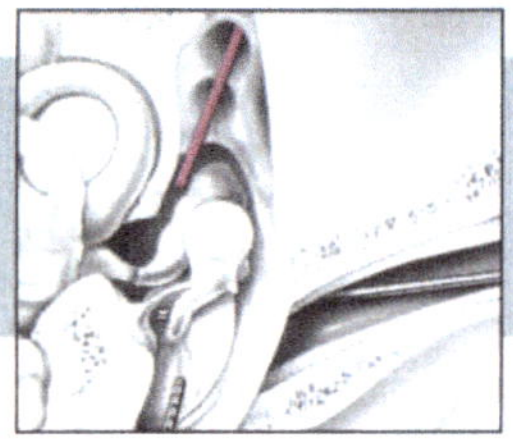

C 1.5.1 Incudomallear joint and chorda tympani

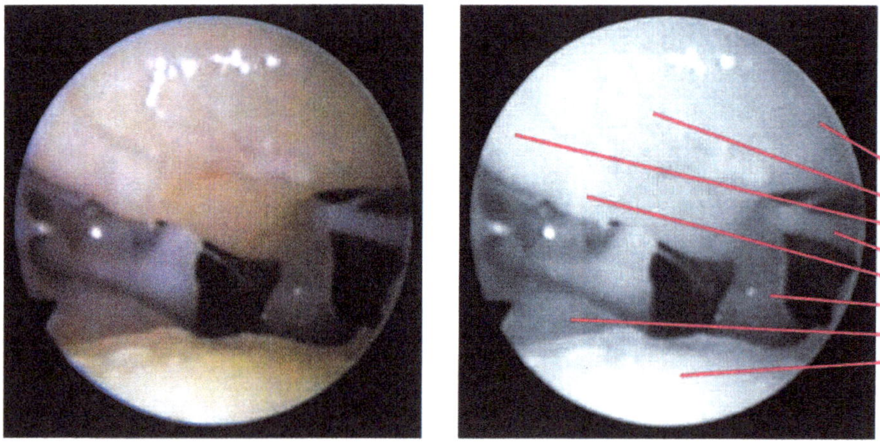

Fig. 142 Scope parallel to tegmen tympani looking down and laterally toward incus (rigid scope, 1.9 mm, 60°, right ear).

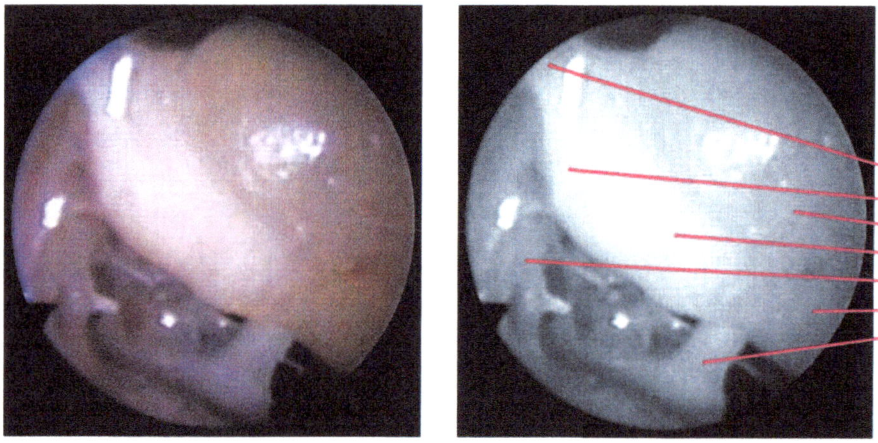

Fig. 143 Advancing the scope brings into view incudomallear joint in the center (rigid scope, 1.9 mm, 60°).

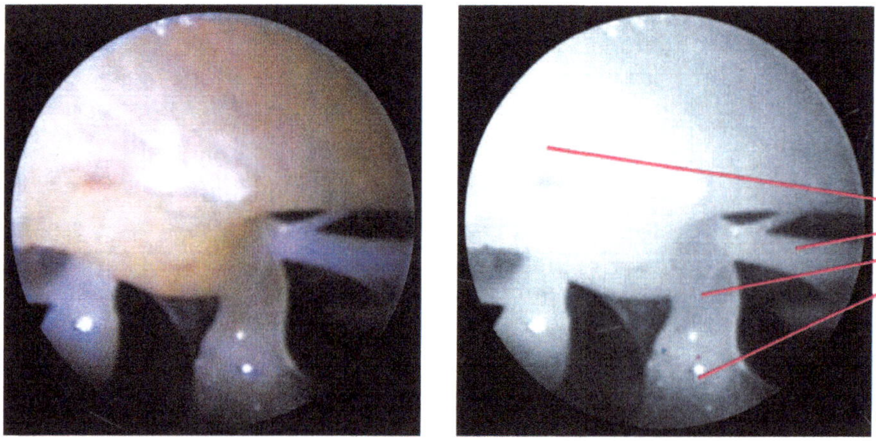

Fig. 144 Tipping the scope downward without rotating it shows the chorda tympani (rigid scope, 1.9 mm, 60°).

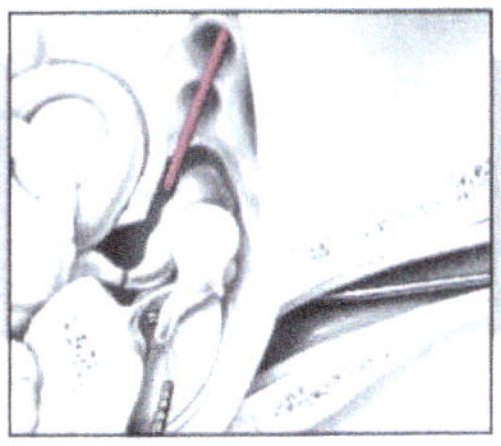

C 1.5.2 Stapes and posterior tympanic isthmus

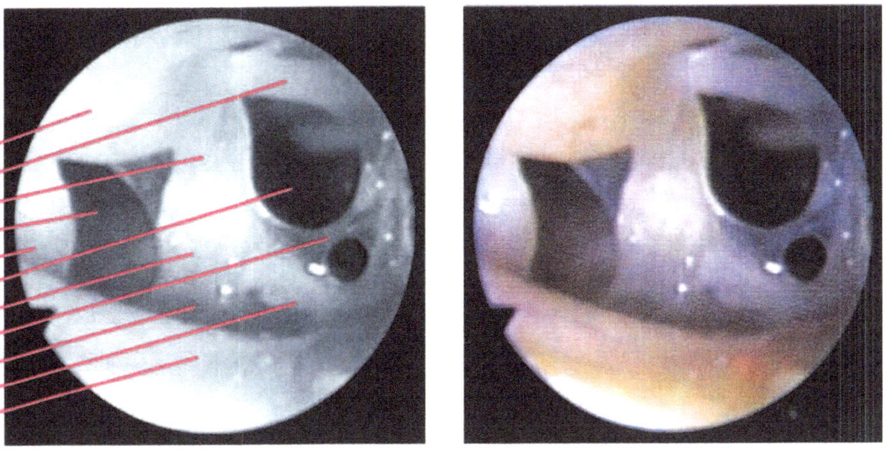

Fig. 145 Continued step-by-step tilting of the scope brings into view the 2 communications of the mesotympanum with the antrum mastoideum (rigid scope, 1.9 mm, 60°, right ear).

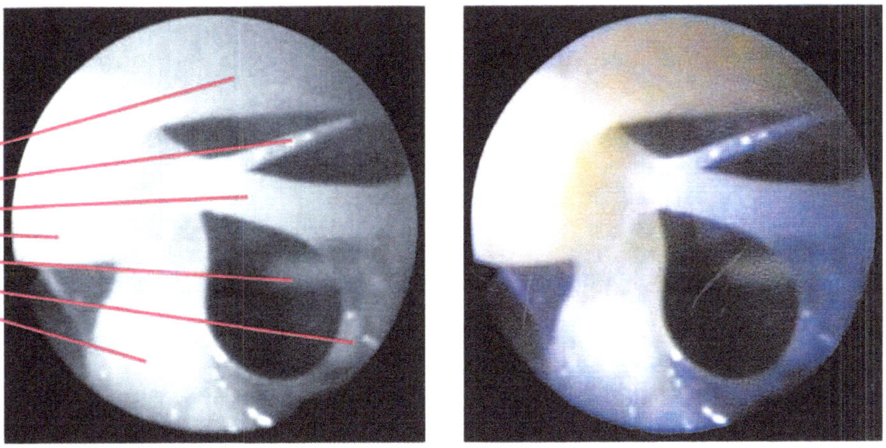

Fig. 146 The posterior tympanic isthmus and the structures bordering it are seen in the center (rigid scope, 1.9 mm, 60°).

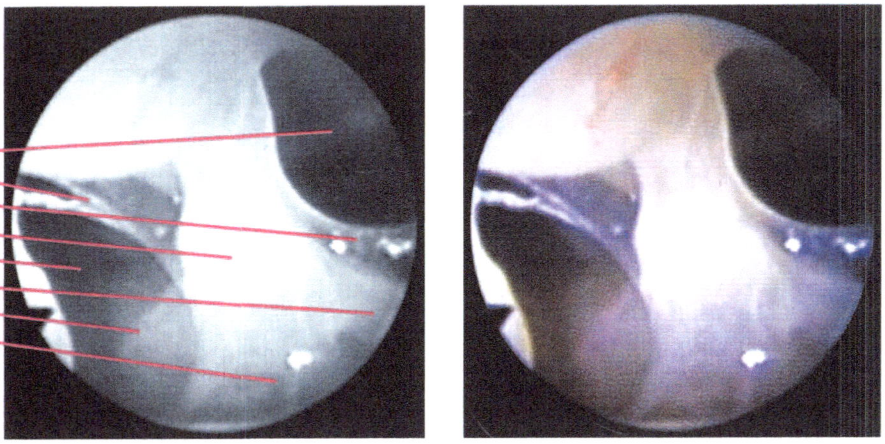

Fig. 147 Moving the scope straight ahead brings the incudostapedial joint into the center of the field of view (rigid scope, 1.9 mm, 30°).

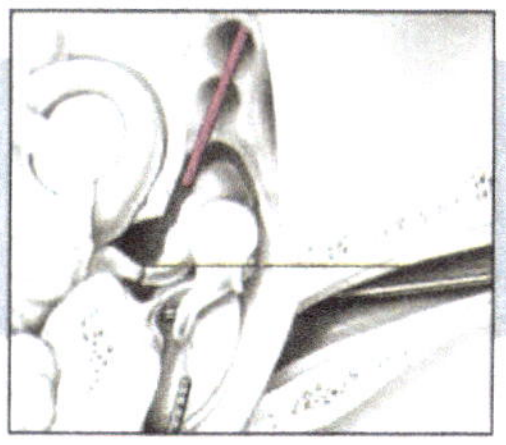

C 2 Transmastoid approach through facial recess

1.9 mm rigid scope

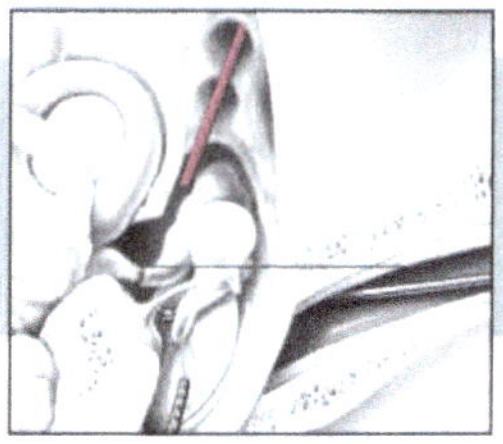

C 2.1 Site of approach

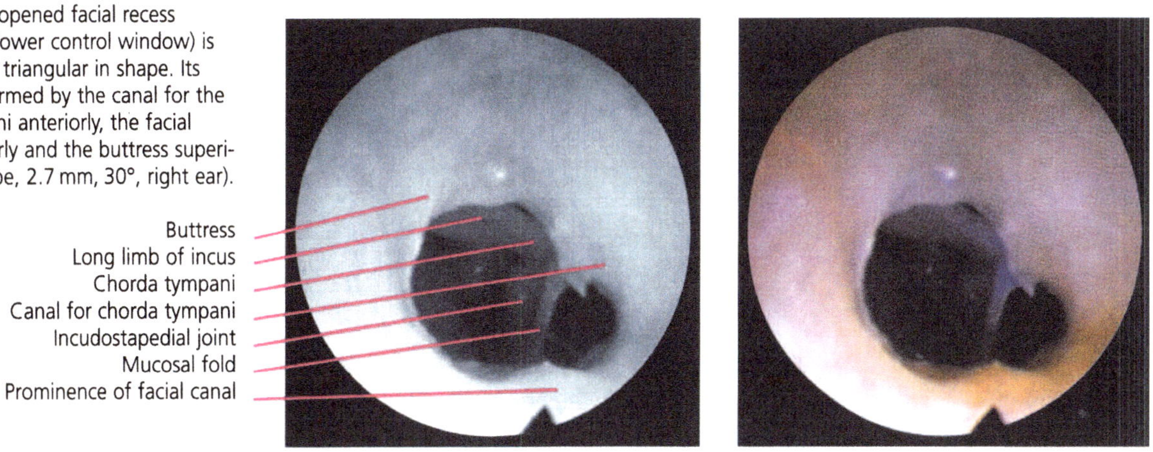

Fig. 148 The opened facial recess (WULLSTEIN's lower control window) is approximately triangular in shape. Its borders are formed by the canal for the chorda tympani anteriorly, the facial canal posteriorly and the buttress superiorly (rigid scope, 2.7 mm, 30°, right ear).

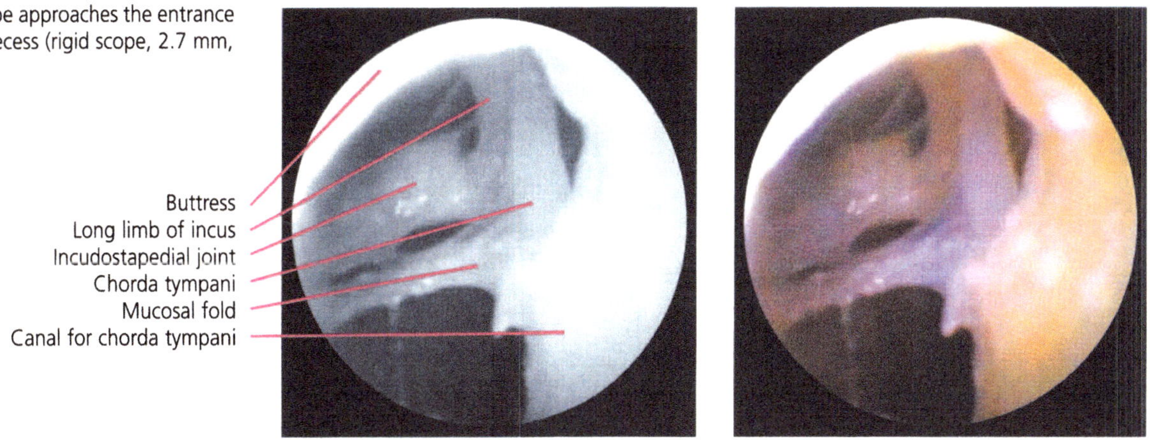

Fig. 149 Scope approaches the entrance to the facial recess (rigid scope, 2.7 mm, 30°).

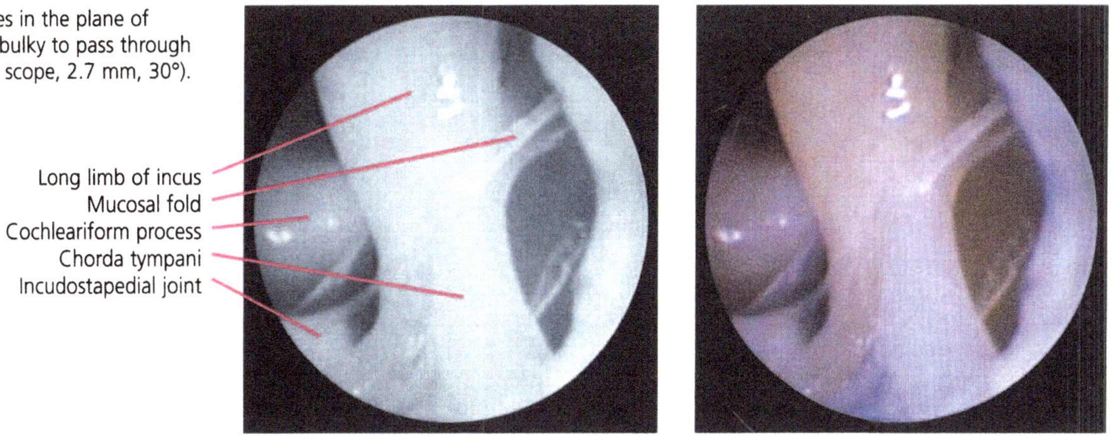

Fig. 150 Scope lies in the plane of access, but is too bulky to pass through the opening (rigid scope, 2.7 mm, 30°).

C 2.2.1 Looking laterally: Chorda tympani and tympanic membrane

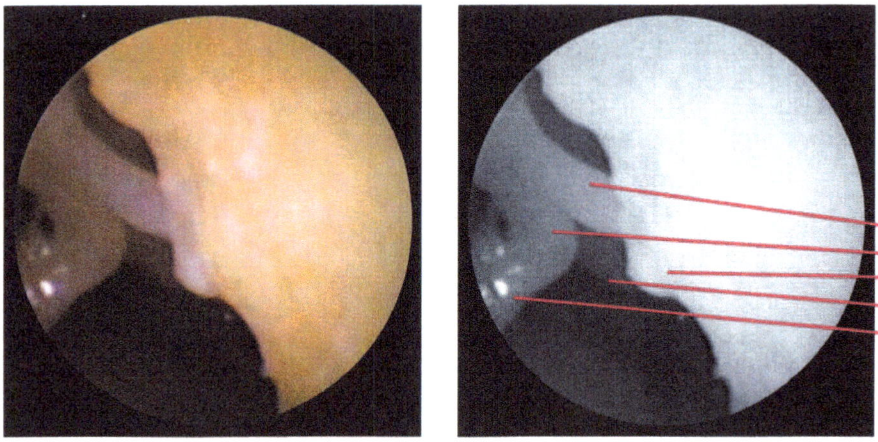

Fig. 151 As the scope passes through the facial recess, the chorda tympani is seen (rigid scope, 1.9 mm, 30°, right ear).

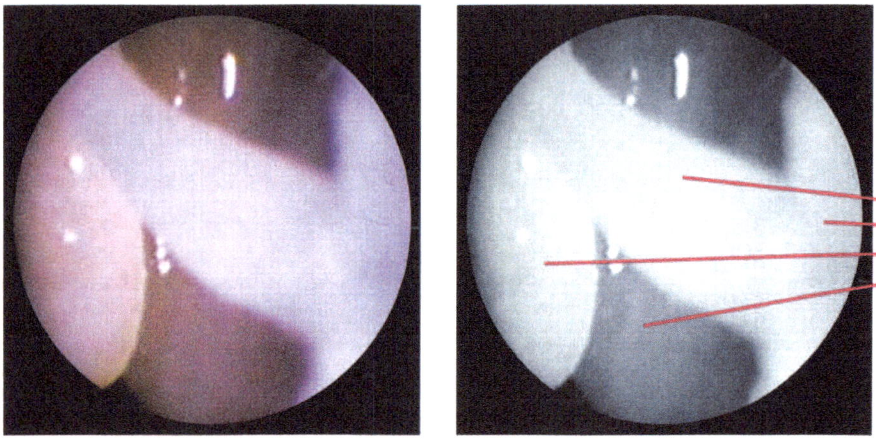

Fig. 152 The scope approaches the site at which the chorda tympani enters or exits its canal (rigid scope, 1.9 mm, 30°).

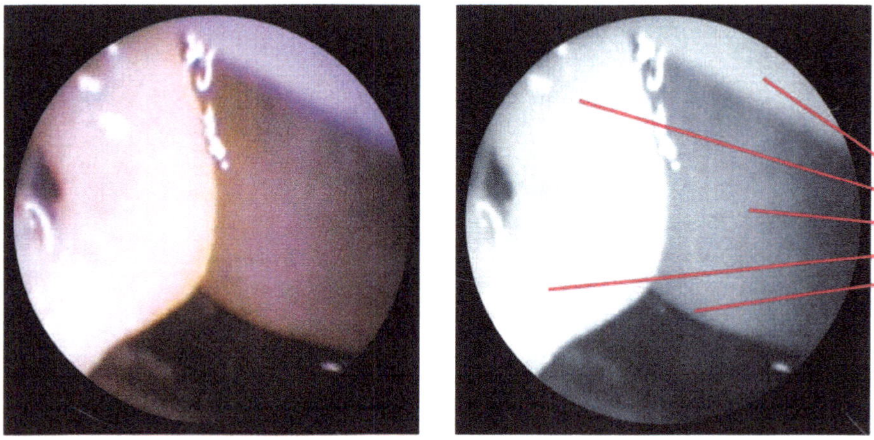

Fig. 153 As the scope is further advanced underneath the chorda tympani, the inner surface of the tympanic membrane comes into view (rigid scope, 1.9 mm, 30°).

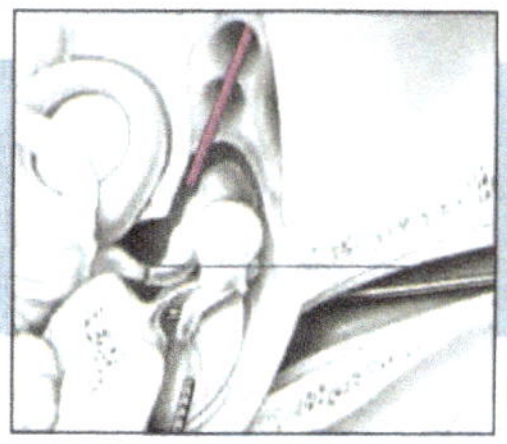

C 2.2.2 Looking medially: Pyramidal eminence

Fig. 154 The pyramidal eminence is seen in the center (rigid scope, 1.9 mm, 30°, right ear).

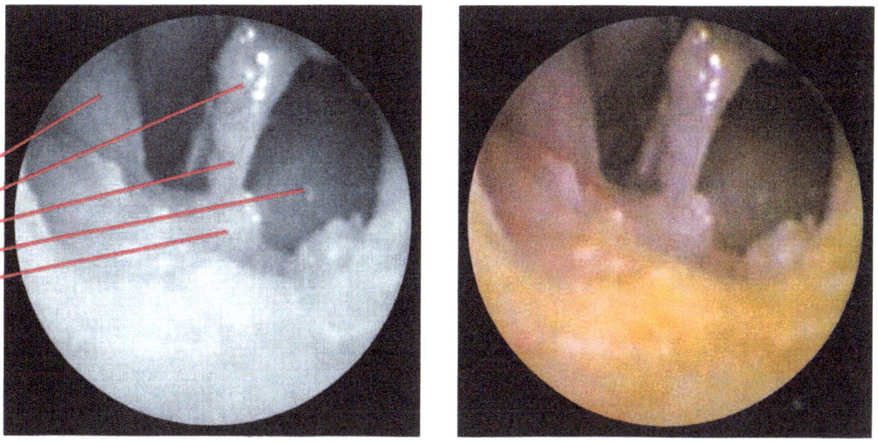

Fig. 155 Along the undersurface of the tendon of the stapedius the scope looks at the promontory and the round window niche (rigid scope, 1.9 mm, 30°).

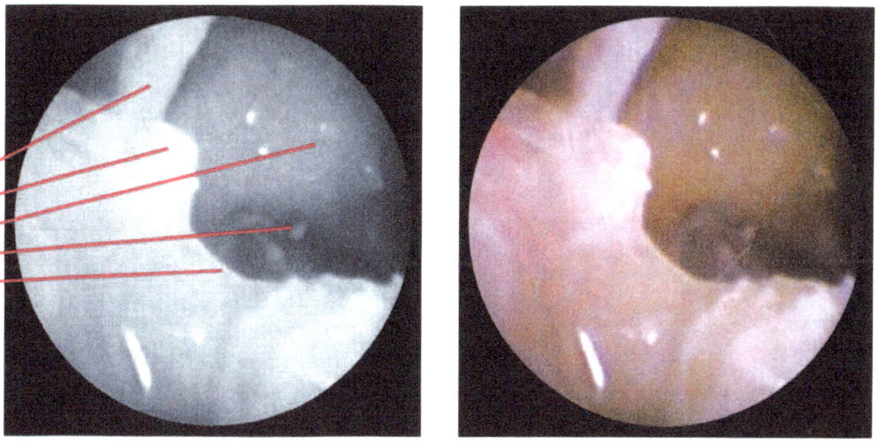

Fig. 156 Along the upper surface of the tendon of the stapedius the scope looks at the facial canal (rigid scope, 1.9 mm, 30°).

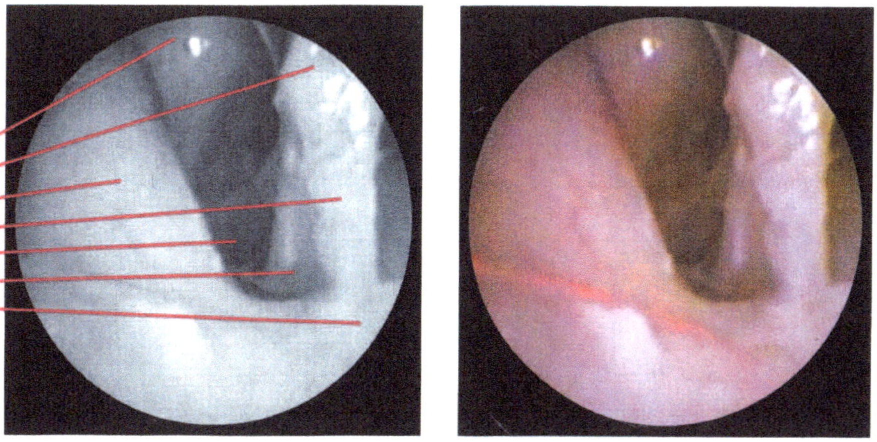

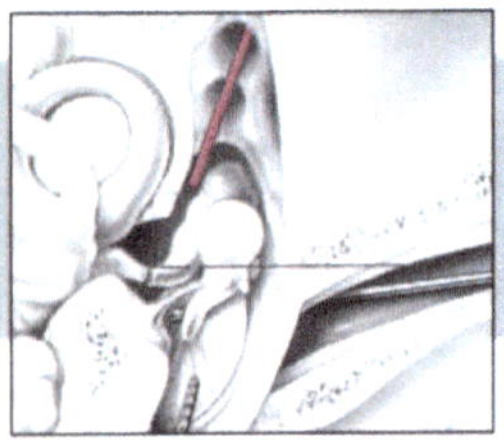

C 2.2.3 Looking up and medially: Facial canal

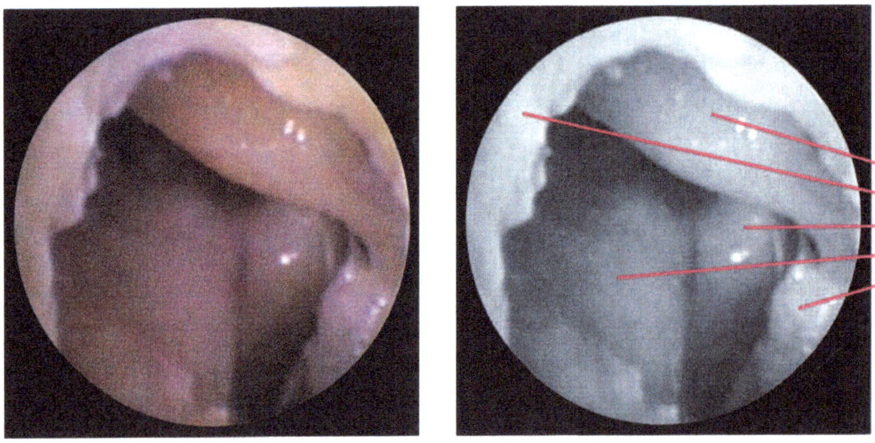

Fig. 157 With the scope at the entry into the facial recess and directed upward an overview of the posterior tympanic isthmus is obtained (rigid scope, 1.9 mm, 30°, right ear).

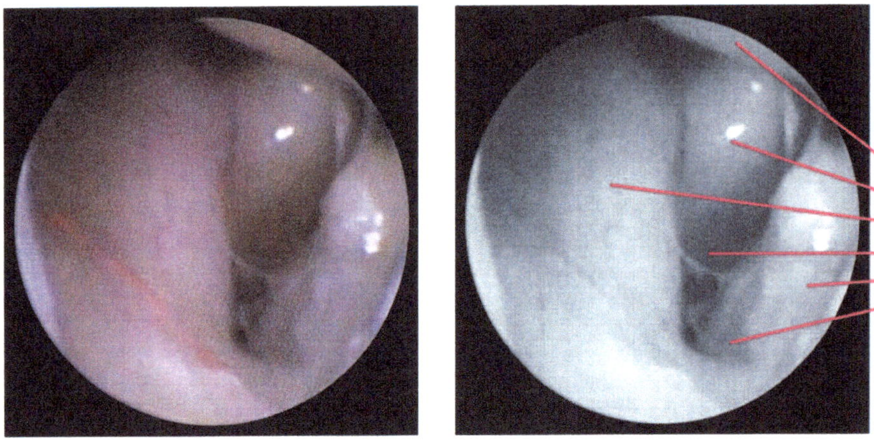

Fig. 158 As the scope is further advanced, the upper border of the oval window niche is seen (rigid scope, 1.9 mm, 30°).

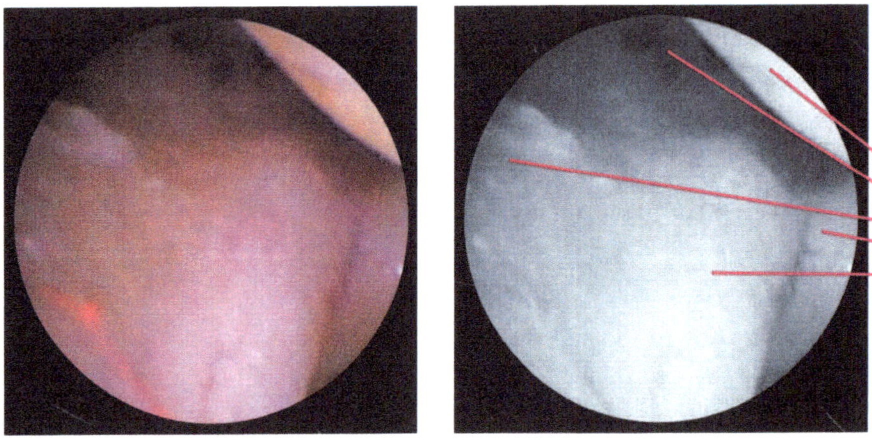

Fig. 159 As the scope comes as close to the facial canal as possible, a view into the epitympanic recess is gained (rigid scope, 1.9 mm, 30°).

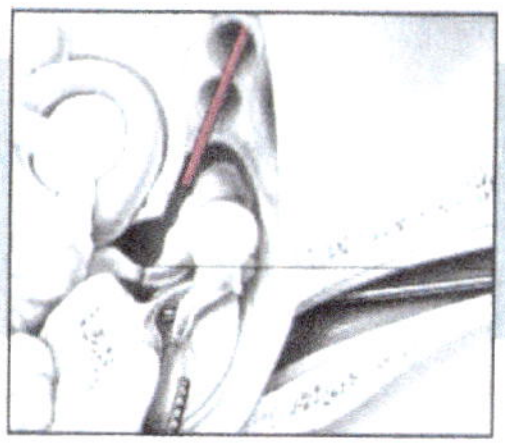

C 2.2.4 Looking down: Promontory and round window niche

Fig. 160 Promontory and round window niche are seen in the center (rigid scope, 1.9 mm, 30°, right ear).

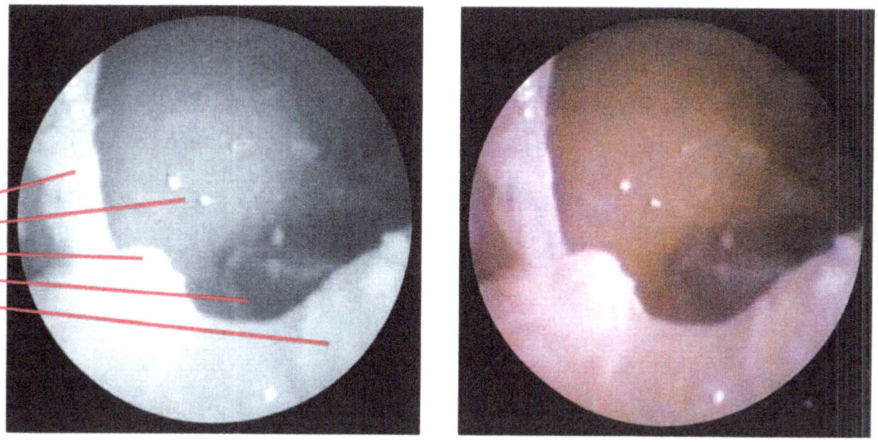

Fig. 161 With the scope advanced and slightly rotated counterclockwise the round window niche is seen (rigid scope, 1.9 mm, 30°).

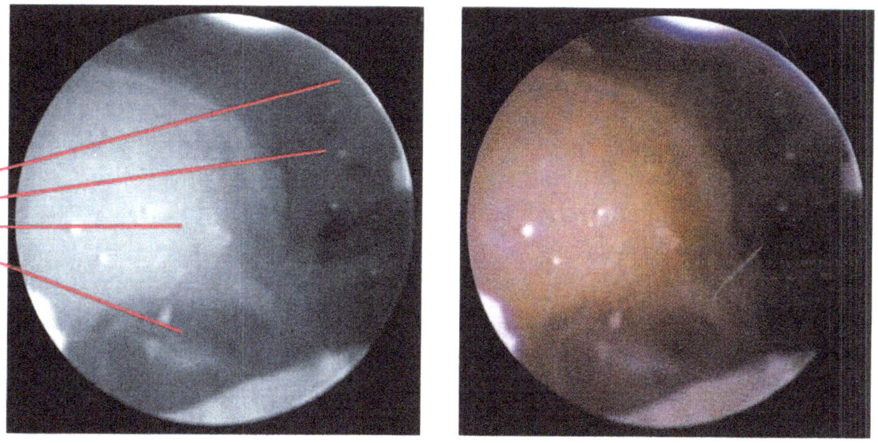

Fig. 162 Continued counterclockwise rotation brings into view the hypotympanum in the center (rigid scope, 1.9 mm, 30°).

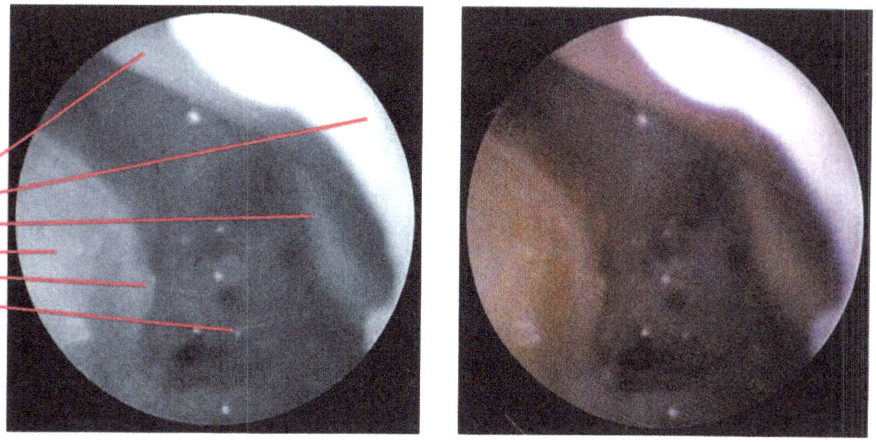

C 2.2.5 Looking down and laterally: Hypotympanum

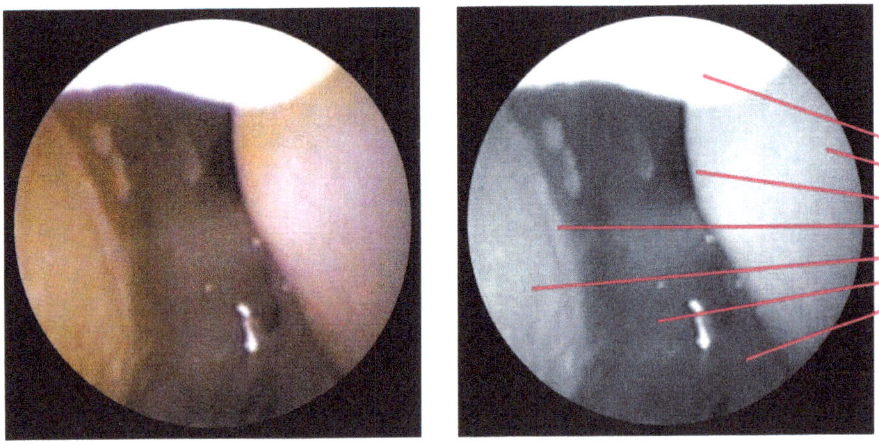

Fig. 163 Scope lies underneath the incudostapedial joint and looks at the inner surface of the tympanic membrane (rigid scope, 1.9 mm, 30°, right ear).

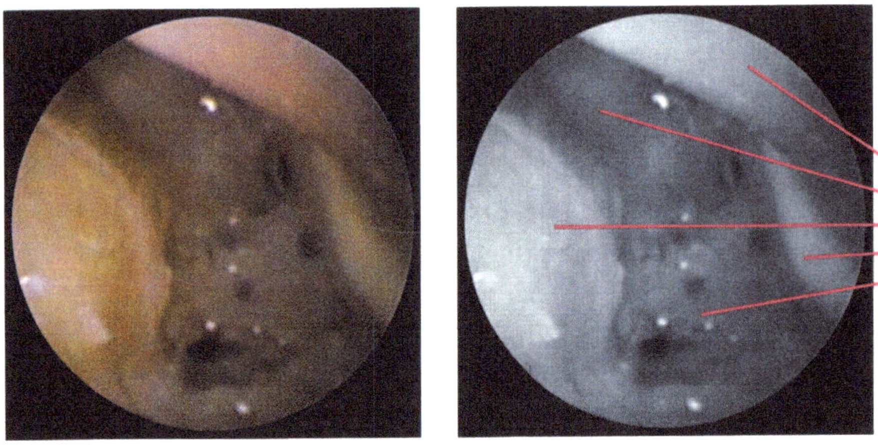

Fig. 164 Rotated clockwise the scope looks at the hypotympanum between the tympanic membrane and the promontory (rigid scope, 1.9 mm, 30°).

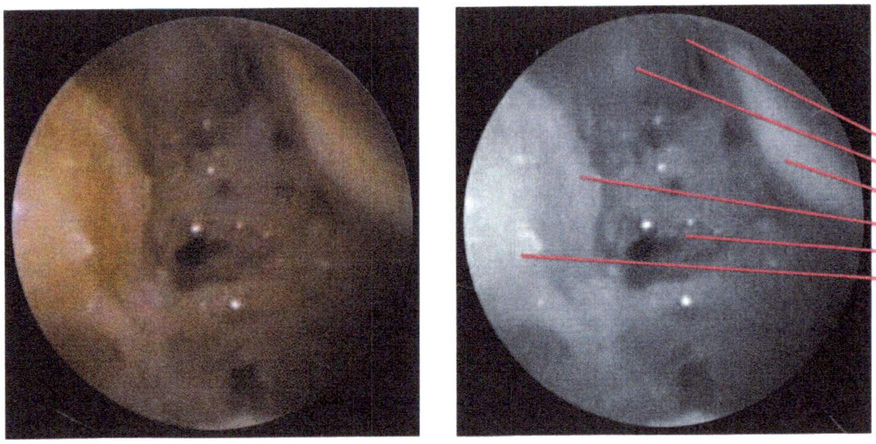

Fig. 165 Advanced as far as possible the scope sees the protympanum (rigid scope, 1.9 mm, 30°).

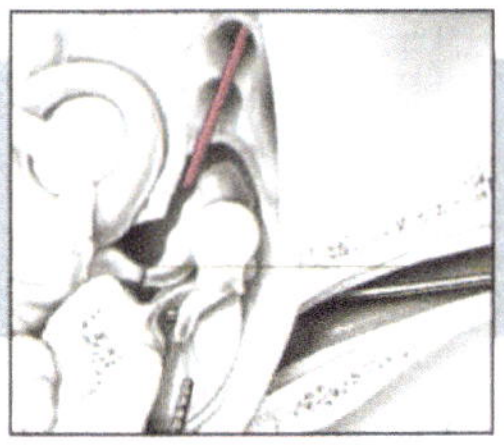

C 2.3.1 Looking down and medially: Round window niche

Fig. 166 With the scope placed as in Fig. 171, but rotated clockwise the posterior hypotympanum is seen between the promontory and the tympanic ring (rigid scope, 1.9 mm, 60°, right ear).

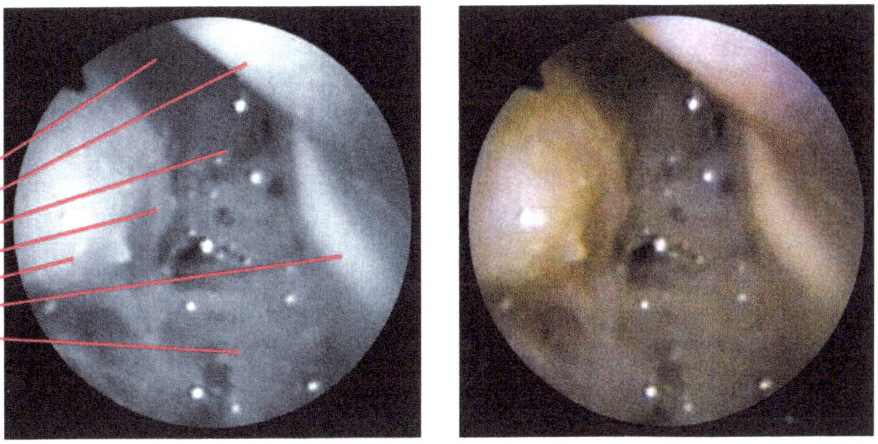

Fig. 167 Rotating the scope medially and looking down the round window niche is seen (rigid scope, 1.9 mm, 60°).

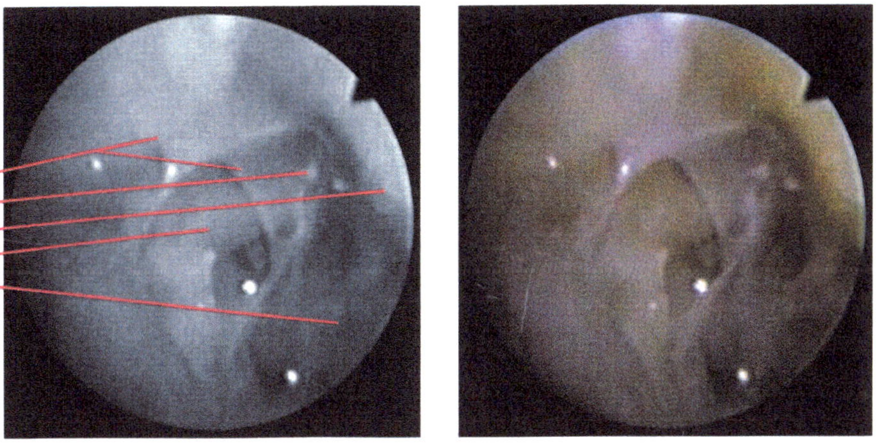

Fig. 168 With the scope rotated as in Fig. 173, but tilted upward the pyramidal eminence and the tendon of stapedius come into view (rigid scope, 1.9 mm, 60°).

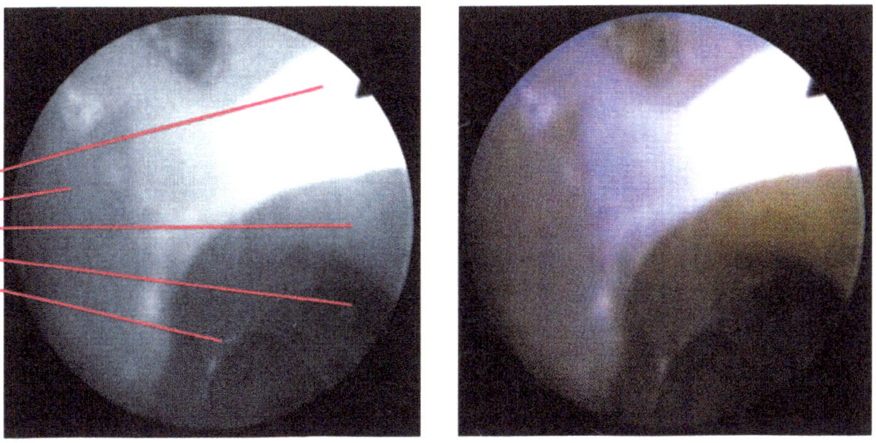

C 2.3.2 Looking down and laterally: Tympanic ring and tympanic membrane

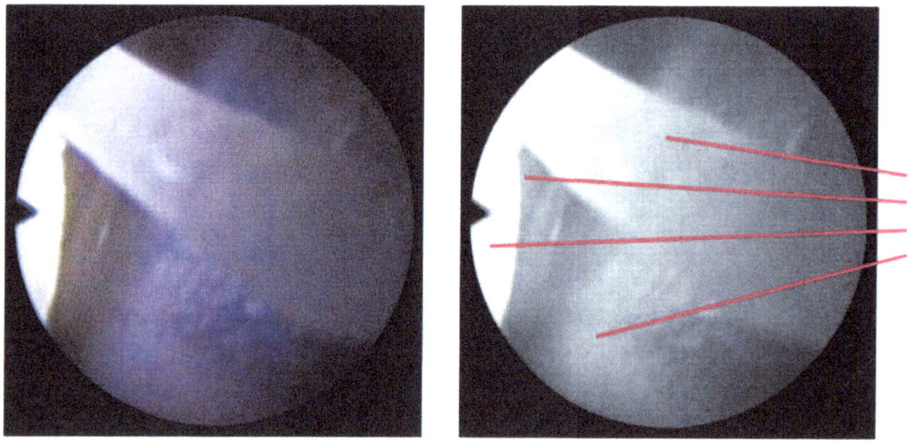

Fig. 169 Direct view of the chorda tympani (rigid scope, 1.9 mm, 60°, right ear).

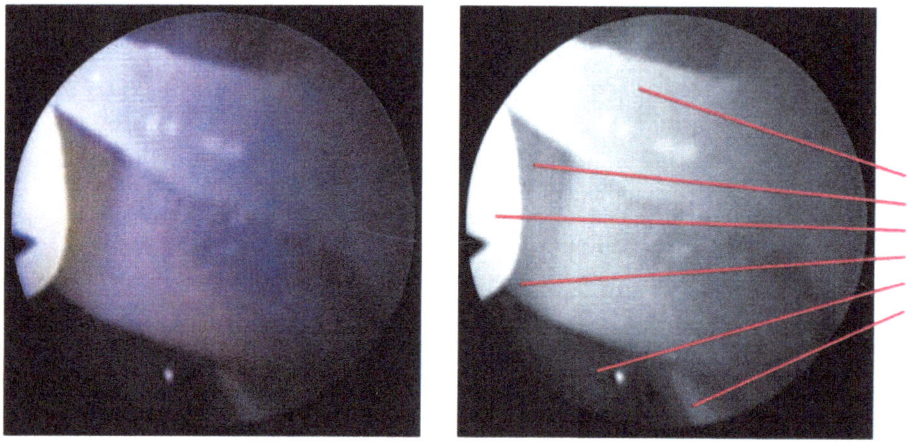

Fig. 170 As the scope is advanced and slightly tipped downward, the umbo and the inferior posterior quadrant of the tympanic cavity come into its field of view (rigid scope, 1.9 mm, 60°).

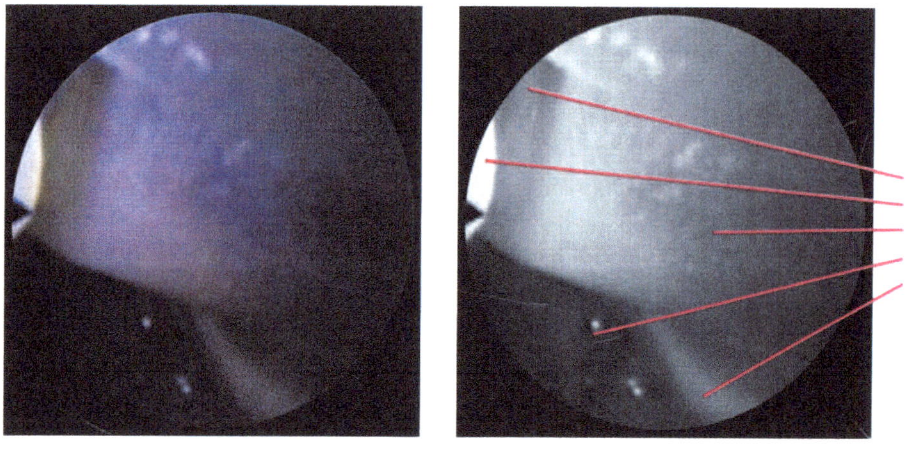

Fig. 171 Tipping the scope further downward shows the tympanic ring (rigid scope, 1.9 mm, 60°).

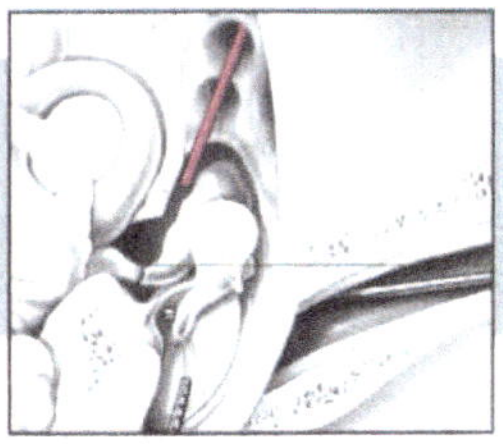

C 2.3.3 Looking up and laterally: PRUSSAK's space

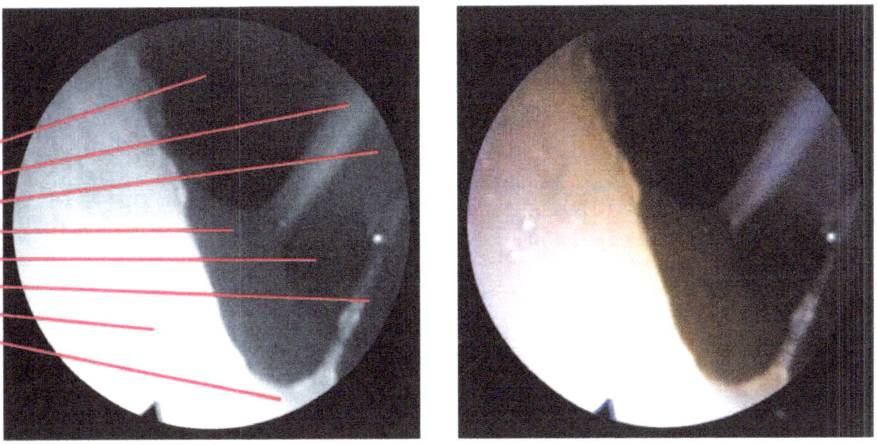

Fig. 172 The scope is looking steeply upward and laterally into the epitympanic recess (rigid scope, 1.9 mm, 60°, right ear).

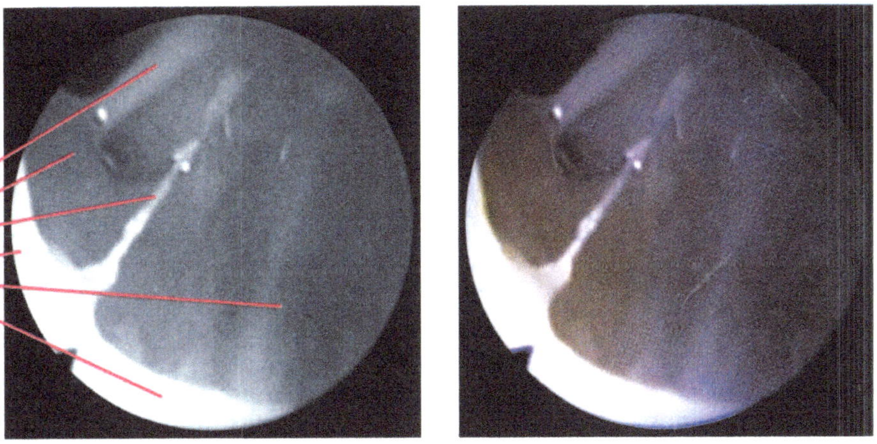

Fig. 173 The scope lies in the plane of the chorda tympani, is slightly rotated clockwise and looks at the posterior superior quadrant of the tympanic membrane (rigid scope, 1.9 mm, 60°).

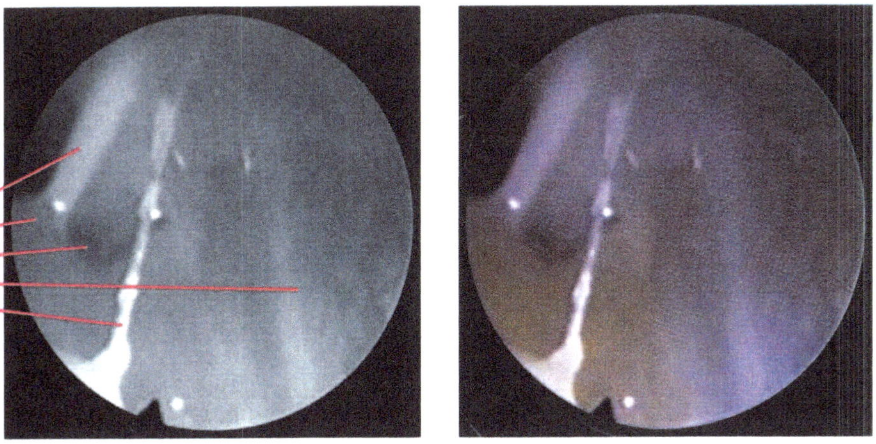

Fig. 174 Further advancement of the scope without rotation brings into view the posterior opening of PRUSSAK's space (rigid scope, 1.9 mm, 30°).

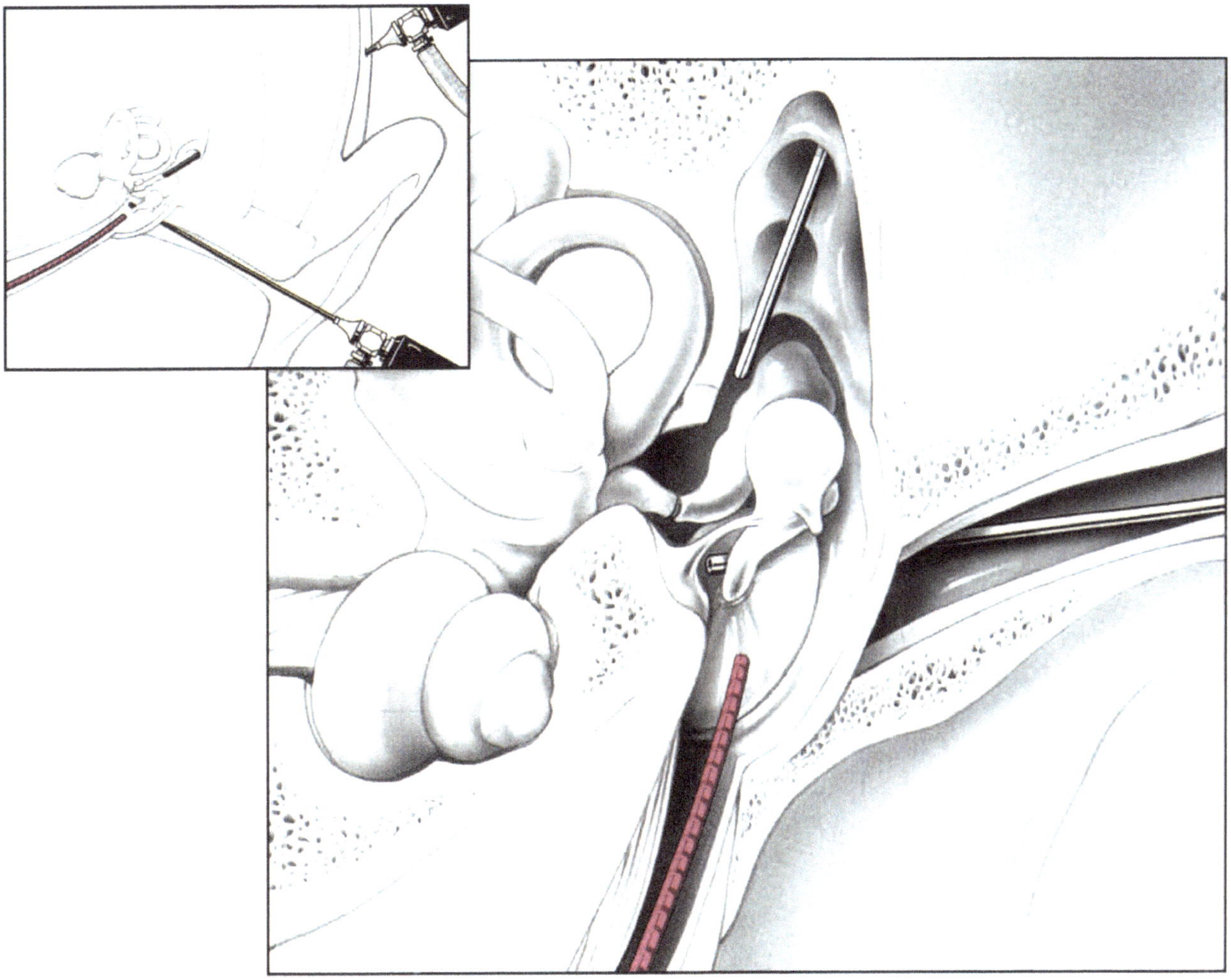

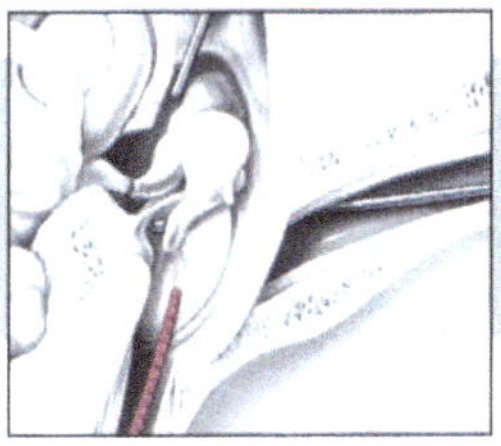

Transtubal approach to the tympanic cavity

The transtubal approach is the only route of access to the middle ear through a natural channel. Along this channel which is formed by the pharyngeal opening of the EUSTACHIAN tube, the EUSTACHIAN tube and the tympanic opening of the EUSTACHIAN tube the scope does not encounter any obstacles so that the structures it passes remain intact. The size of the scope is limited by the width of the isthmus, which will usually not accept scopes larger than 1 mm.

For introducing the scope a tubal catheter (EUSTACHIAN catheter) is used as a "guide". Its placement is ideally monitored through a rigid 4 mm 70° scope in the contralateral nasal cavity.

Once in place, the scope provides an anteroposterior view of the walls of the tympanic cavity (except the anterior wall; "keyhole" effect!) and of the ossicles.

The type of scope used, i.e. flexible but not steerable or flexible and steerable, determines what can be seen.

(1) Flexible non-steerable scopes:

Bony trabeculae of variable size and shape projecting into the lumen of the protympanum may deflect the tip of the scope. Mesotympanic structures may also obstruct its progress.

a. When the scope is advanced above the tendon of the tensor tympani muscle, it impinges upon the mucosal fold. Perforation of this fold gives free access to the epitympanum.

b. When the scope is advanced below the tendon, the mesotympanum can be readily explored. (This is the best possible case.)

c. When the scope cannot be advanced beyond the tendon of the tensor tympani, but impinges upon it, no further exploration is possible with non-steerable instruments.

(2) Flexible steerable scopes:

The "obstacles" described above can easily be bypassed by steerable scopes. These also provide a view into the hypotympanum. But controlled steerability has its price: At a given outer diameter the number of image points is smaller than with non-steerable scopes, because part of the cross-sectional area is occupied by the steering mechanism. As a result, the images are of poorer quality.

What type of scope was chosen for the images in this Section depended on the underlying anatomy. For short and wide tubes flexible non-steerable scopes can be used, because their movements can be controlled to a certain extent by the EUSTACHIAN catheter. Flexible non-steerable scopes were used for Figs. 178-192.

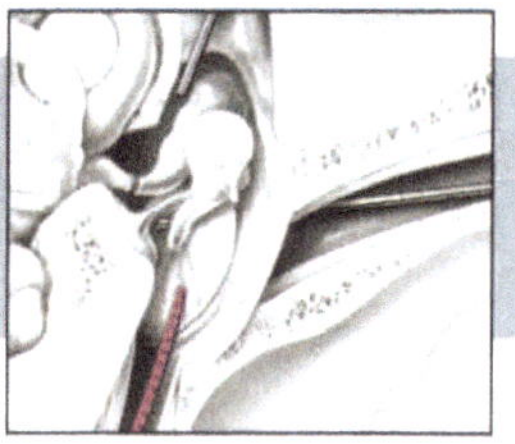

For very narrow tubes steerable scopes are indispensable. These were used for all other illustrations.

The overviews with the transtubal scope shown in place were obtained with a rigid 1.9 mm 30° scope in the external auditory canal. This was either inserted through an anterior inferior (Figs. 178, 181, 202) or posterior inferior quadrant (Figs. 184, 187, 190) paracentesis or by reflecting a tympanomeatal flap (Figs. 208, 214, 217, 220, 223, 226, 229, 232).

Landmarks for orientation – from the top down and from medial to lateral

As a rule, the tendon of the tensor tympani is the first tympanic structure seen. It is easily recognized by its horizontal course and its characteristic relation to the handle of the malleus, which comes into view peripherally at the same time. These 2 structures are arranged at an angle of about 75° in a "figure-of-7" shape on the left side and its mirror image on the right. When the scope is advanced, the incudostapedial joint is the next characteristic landmark. While the anterior limb of the stapes is more or less horizontal, the long limb of the incus courses from above and medial to below and lateral and looks like the upside-down counterpart of the handle of the malleus. These 2 structures are also arranged at an angle of 75° in an upside-down "figure-of-7" shape on the right side and its mirror image on the left.

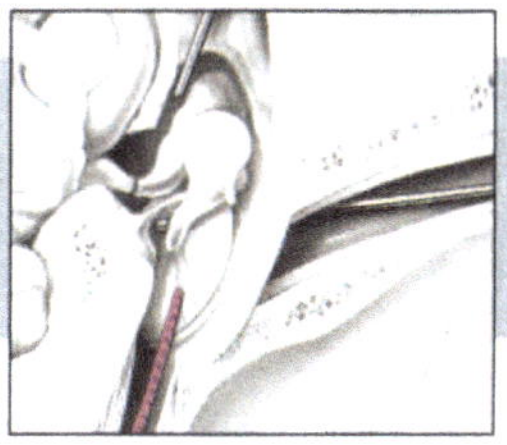

Transtubal approach to the tympanic cavity

0.8 mm flexible steerable scope

D 1.1 Site of approach

Unflected scope on its way from tubal opening to antrum
(not always possible)

D 1.2.1 View from tympanic opening of EUSTACHIAN tube into mesotympanum

D 1.2.2 Scope kept straight to approach mesotympanic structures

D 1.2.3 Scope advanced into central mesotympanum

D 1.2.4 Scope crosses over stapes (medial to incus) and looks into mesotympanum

D 1.2.5 Scope between long limb of incus and facial canal:
Aditus ad antrum

Inspection of tympanic membrane

D 1.3.1 Scope in mesotympanum bent laterally:
Lateral wall of tympanic cavity

D 1.3.2 Scope in mesotympanum bent laterally and upward:
PRUSSAK's space

D 1.3.3 Scope in mesotympanum bent laterally and downward:
Hypotympanum

Inspection of epitympanum

D 1.4.1 Scope in protympanum bent upward:
Mucosal fold of tensor tympani

D 1.4.2 Scope advanced:
Perforation of mucosal fold of tensor tympani

D 1.4.3 Scope above tendon of tensor tympani:
Incudomallear joint

D 1.4.4 Scope medial to incudomallear joint:
Posterior epitympanic recess

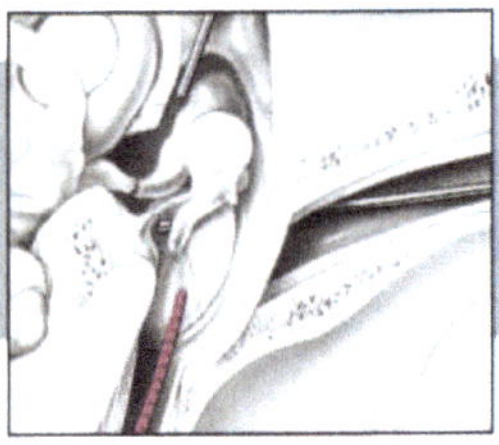

Inspection of antrum

D 1.5.1 **Scope below tendon of tensor tympani: Incudostapedial joint**

D 1.5.2 **Scope medial to long limb of incus: Facial canal and lateral semicircular canal**

D 1.5.3 **Scope crosses over stapes: Aditus ad antrum**

Inspection of chorda tympani
(crossing tensor tympani lateral to incus)

D 1.6.1 **Scope bent laterally and upward behind handle of malleus: Chorda tympani**

D 1.6.2 **Scope lateral to long limb of incus: Course of chorda tympani**

Inspection of posterior wall
(crossing tendon of tensor tympani and stapes)

D 1.7.1 **Scope in the plane of the handle of malleus: Stapes**

D 1.7.2 **Scope below incudostapedial joint: Posterior wall**

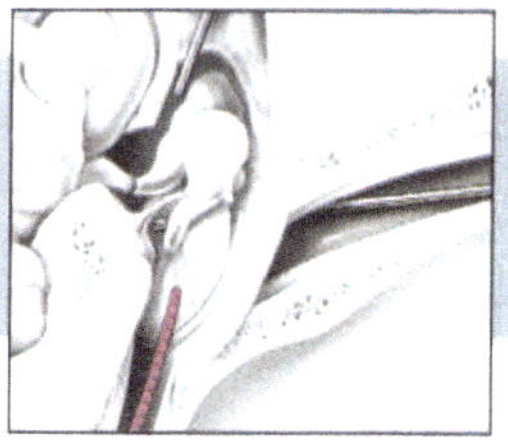

D 1.1 Site of approach

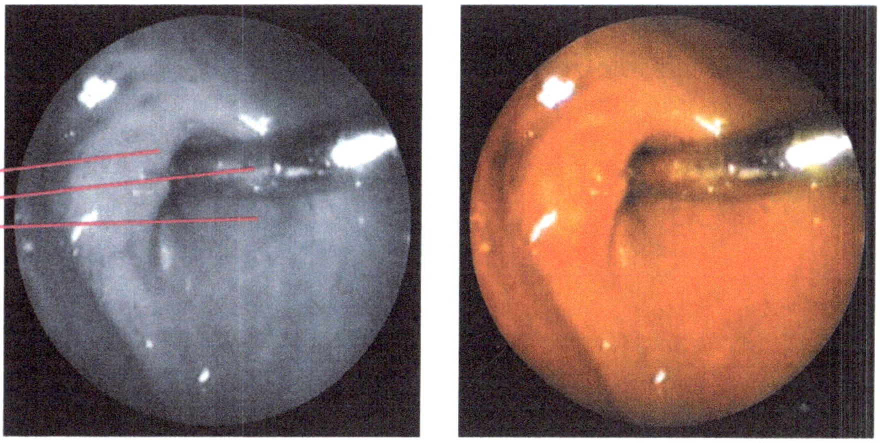

Fig. 175 Transnasal EUSTACHIAN catheter in the left pharyngeal opening of the EUSTACHIAN tube seen through a rigid 4 mm 70° scope in the right nasal cavity.

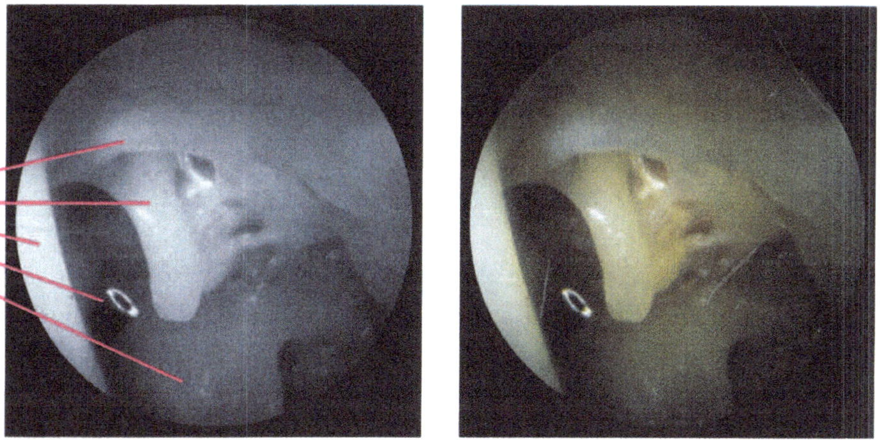

Fig. 176 Transtubal (flexible steerable 0.8 mm) scope in the mesotympanum seen through a rigid transmeatal 2.7 mm 30° scope (left ear).

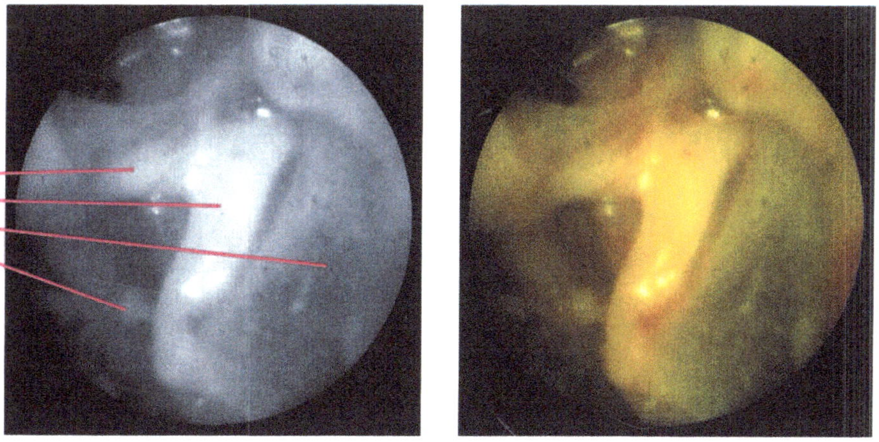

Fig. 177 Overview through (flexible steerable 0.8 mm) transtubal scope from the protympanum toward the mesotympanum (left ear).

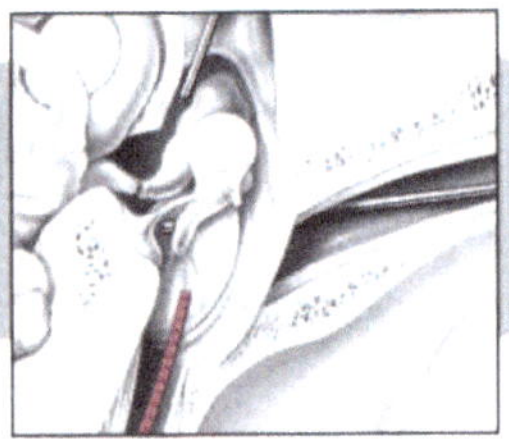

Unflected scope on its way from the tubal opening to the antrum (not always possible)

D 1.2.1 View from tympanic opening of EUSTACHIAN tube into mesotympanum

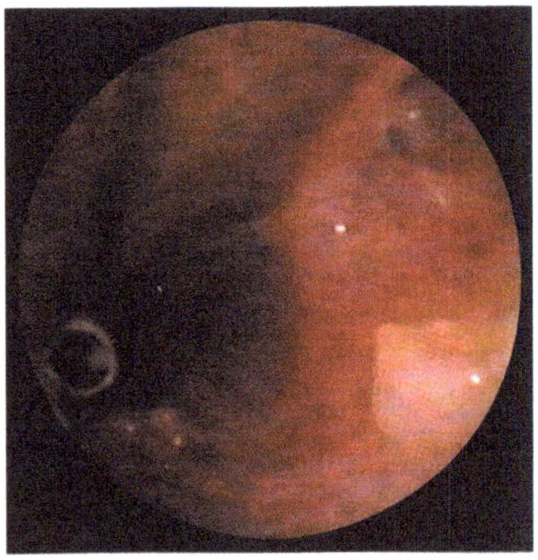

Fig. 178 The transtubal scope lies in the plane of the tympanic opening of the EUSTACHIAN tube and is directed toward the mesotympanum. It is seen through a rigid transtympanic 1.9 mm 30° scope (left ear).

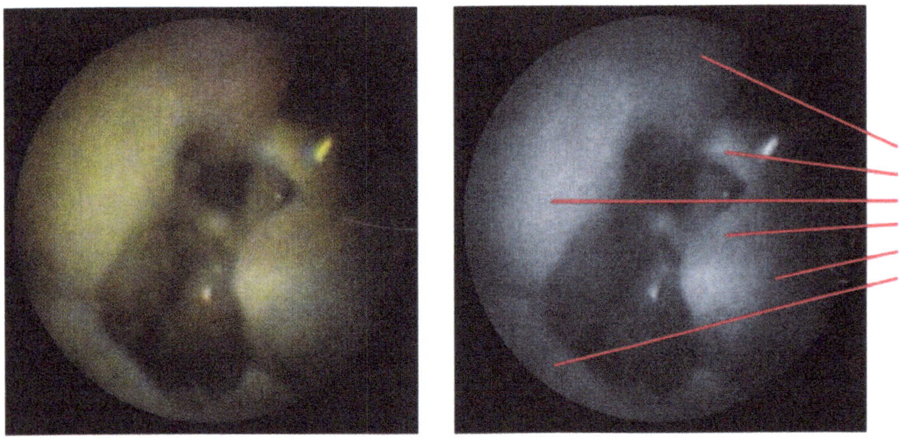

Fig. 179 The scope lies in the protympanum and looks toward the mesotympanum.

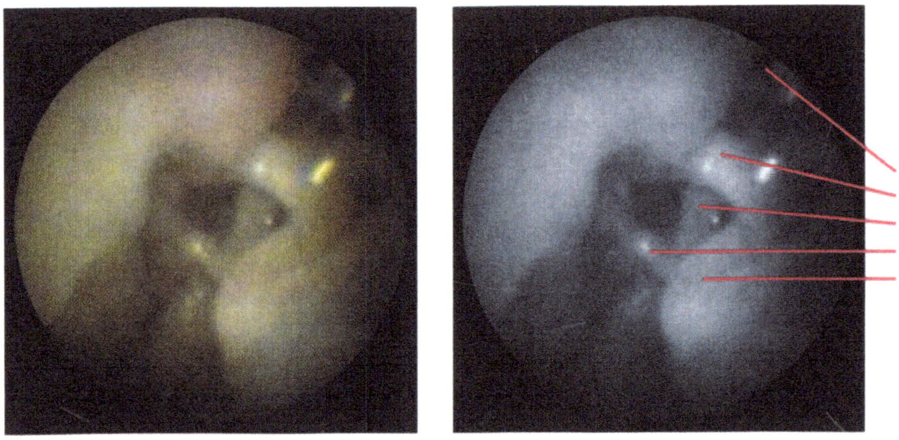

Fig. 180 The scope is 2 mm closer to the mesotympanum than in Fig. 179.

D 1.2.2 Scope kept straight to approach mesotympanic structures

Fig. 181 Without bending it, the transtubal scope was advanced to the anterior border of the tympanic membrane and looks toward the mesotympanum (left ear).

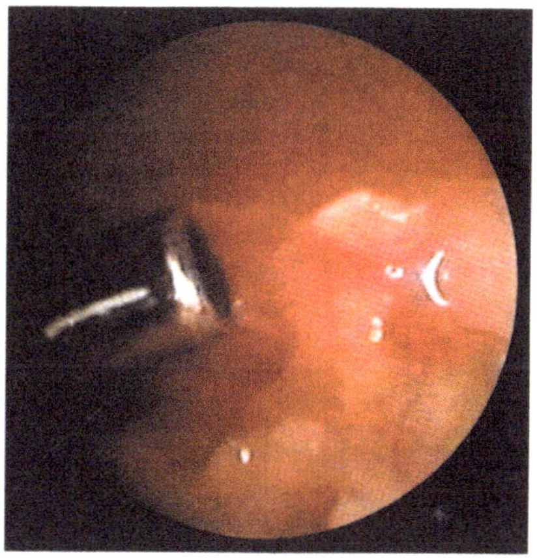

Fig. 182 "Figure-of-7" sign (tendon of tensor tympani and handle of malleus) in the center serves as a landmark for advancing the scope.

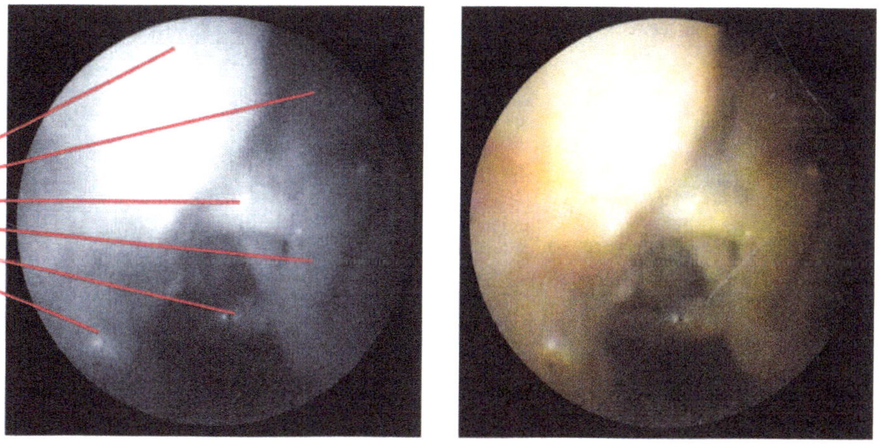

Fig. 183 Still looking in the same direction, the mesotympanic structures are seen below the tendon of the tensor tympani muscle.

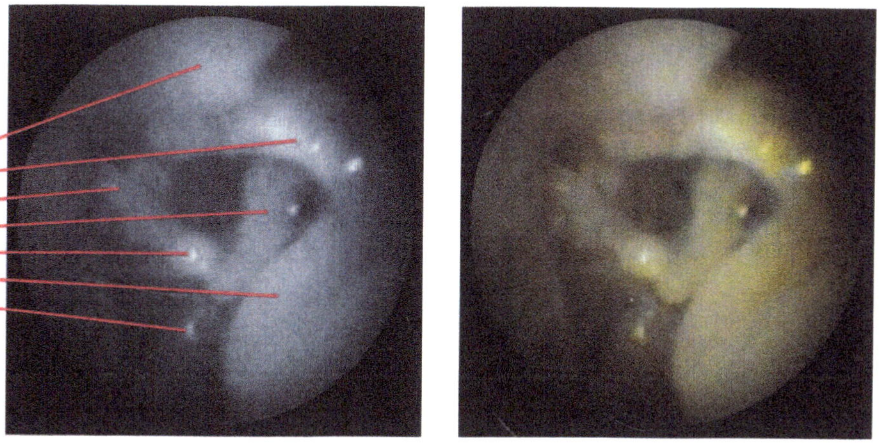

D 1.2.3 Scope advanced into central mesotympanum

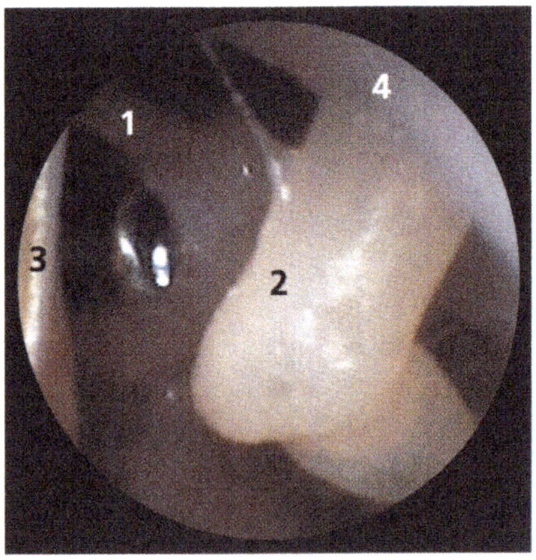

Fig. 184 The transtubal scope lies at the level of the tendon of the tensor tympani (1) and looks toward the antrum mastoideum, (2) long limb of incus, (3) tympanic membrane, (4) chorda tympani (left ear).

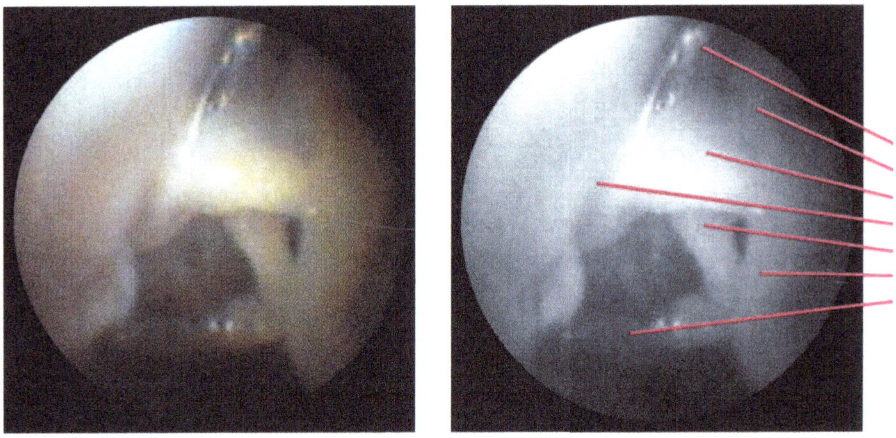

Fig. 185 "Figure-of-7" formed by handle of malleus* and tendon of tensor tympani** clearly stands out as landmark.

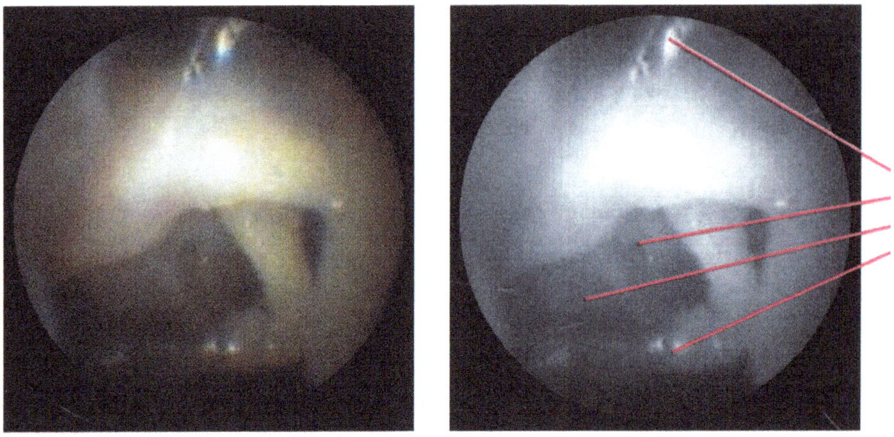

Fig. 186 Scope brought closer for a magnified view of the details shown in Fig. 185.

D 1.2.4 Scope crosses over stapes (medial to incus) and looks into mesotympanum

Fig. 187 Passing underneath the tendon of the tensor tympani, the transtubal scope is advanced straight ahead (i.e. without bending) toward the antrum mastoideum. It comes to lie above the stapes (1) and lateral to the long limb of the incus (2) (left ear).

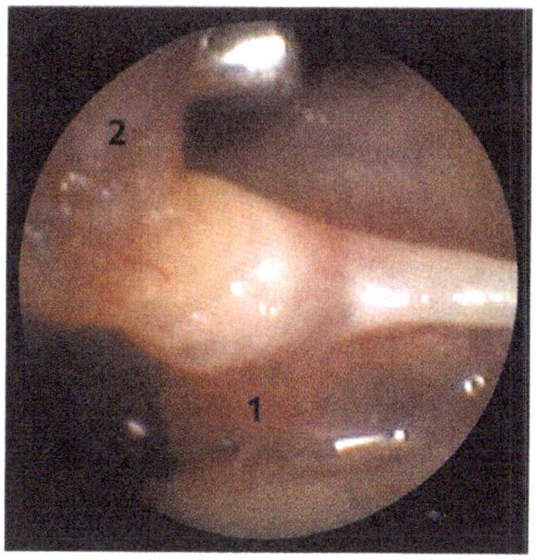

Fig. 188 Upside-down "figure-of-7"

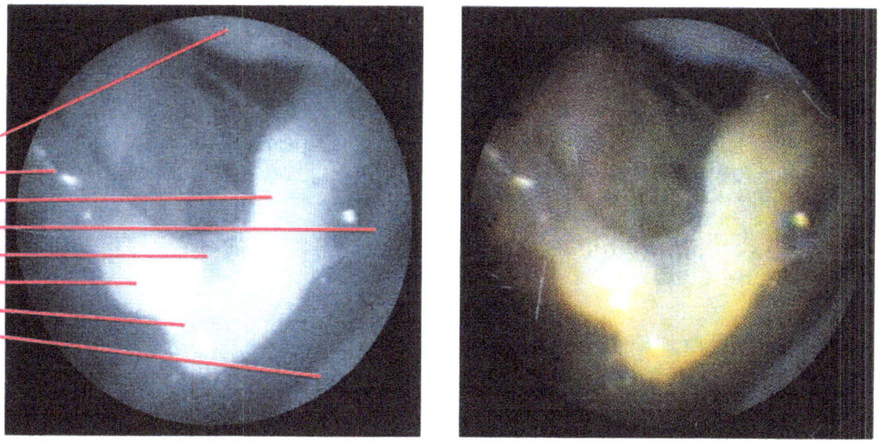

* Upside-down "figure-of-7"

Fig. 189 At 12 o'clock the antrum mastoideum and the structures bordering it come into view.

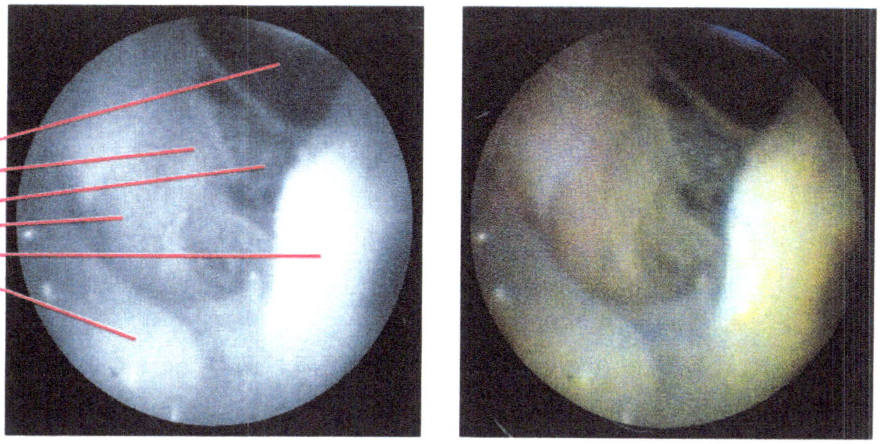

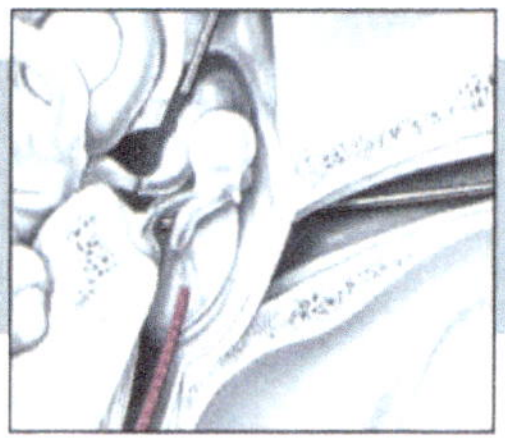

D 1.2.5 Scope between long limb of incus and facial canal: Aditus ad antrum

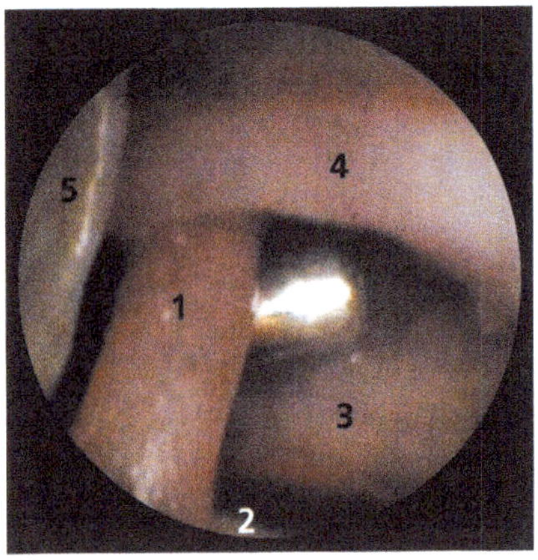

Fig. 190 The scope is advanced straight ahead underneath the tendon of the tensor tympani and comes to lie in the gap between the incus (1), the stapes (2) and the facial canal (3). The chorda tympani (4) and the tympanic membrane (5) are also seen (left ear).

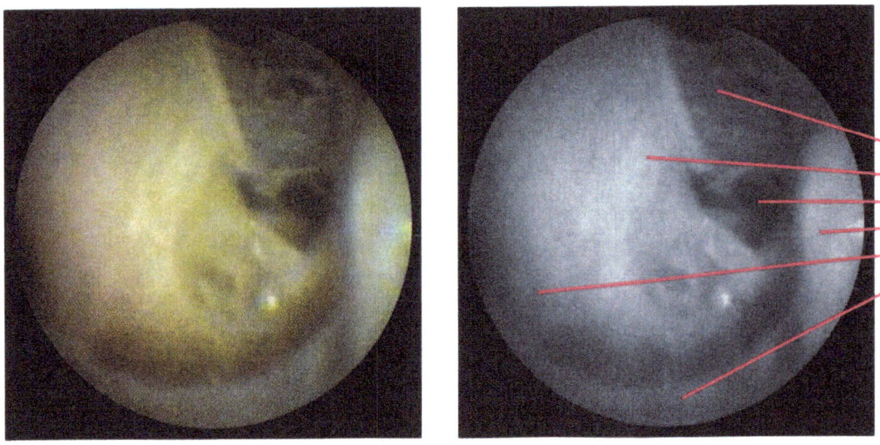

Fig. 191 View past the incus toward the aditus ad antrum.

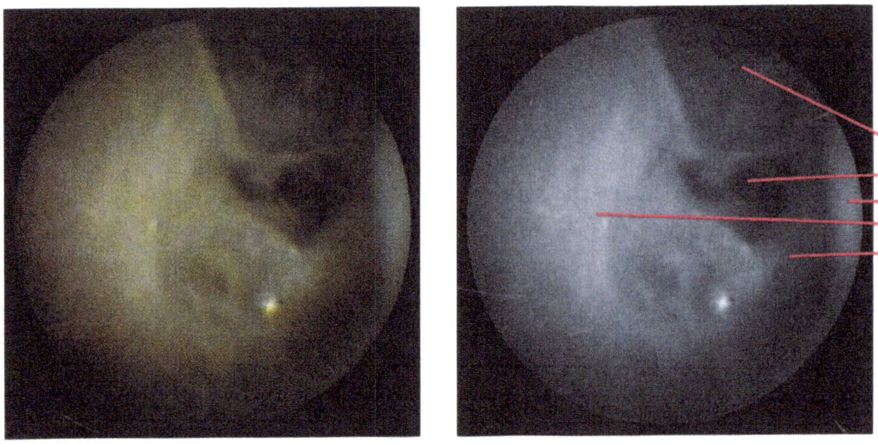

Fig. 192 Scope moved closer to the aditus ad antrum.

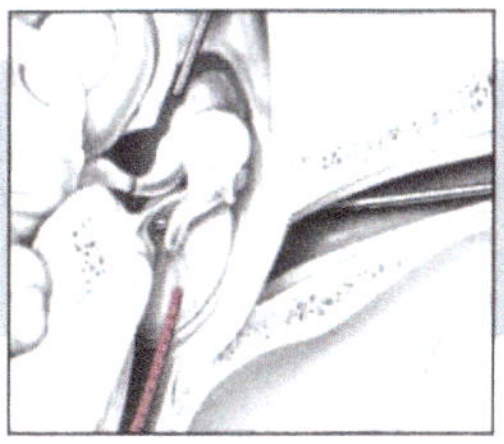

Inspection of tympanic membrane

D 1.3.1 Scope in mesotympanum bent laterally: Lateral wall of tympanic cavity

Fig. 193 Overview of the mesotympanic structures seen through the flexible transtubal scope. Figs. 194-201 show the lateral segment of the tympanic cavity. (1) Tendon of tensor tympani, (2) handle of malleus, (3) long limb of incus, (4) stapes (left ear).

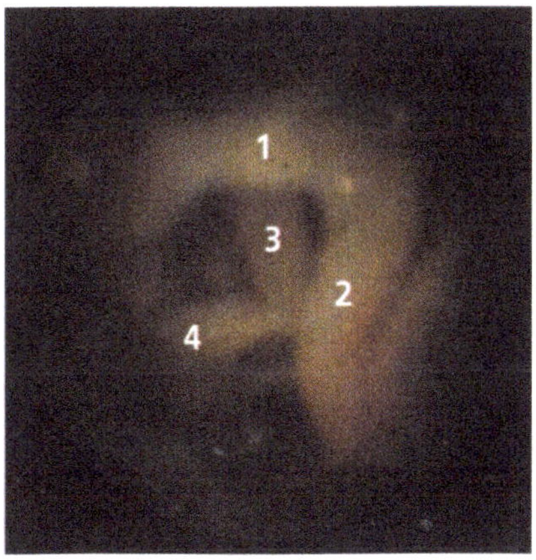

Fig. 194 Lateral sweep with flexible scope.

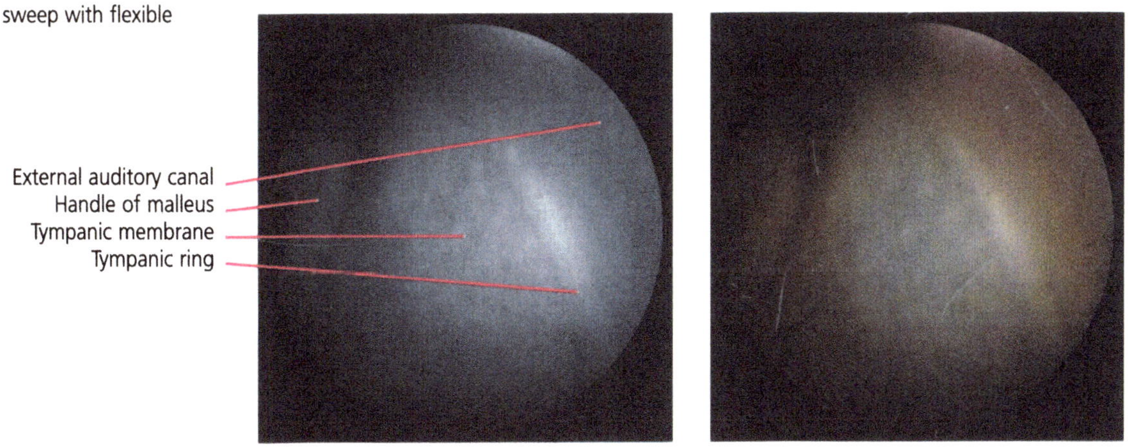

Fig. 195 Scope focused on umbo on the inner side of the tympanic membrane.

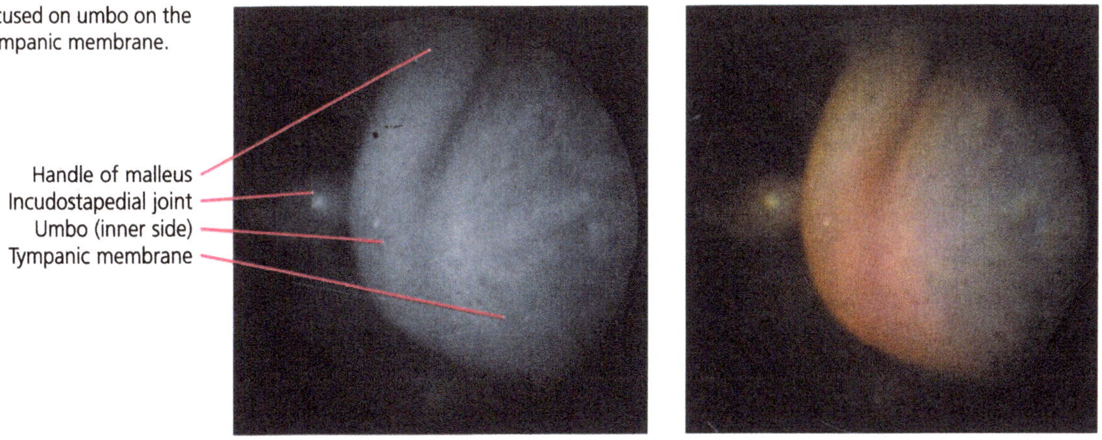

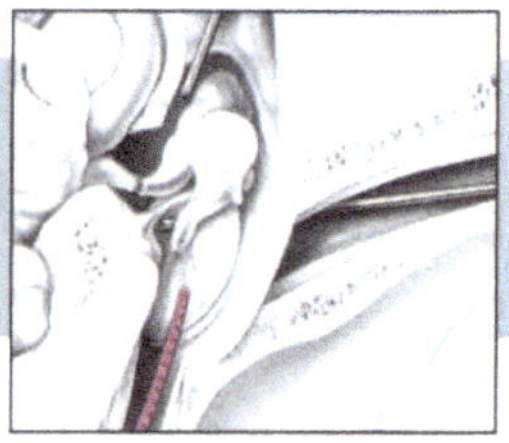

D 1.3.2 Scope in mesotympanum bent laterally and upward: PRUSSAK's space

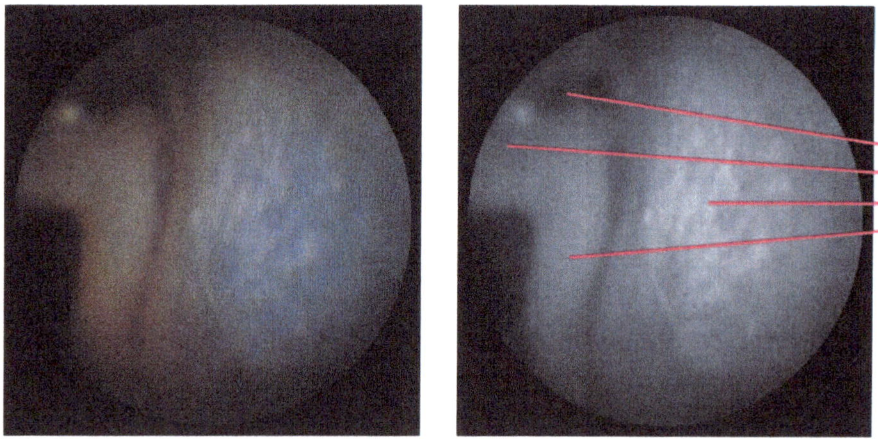

Fig. 196 The anterior superior quadrant of the tympanic membrane is seen (left ear).

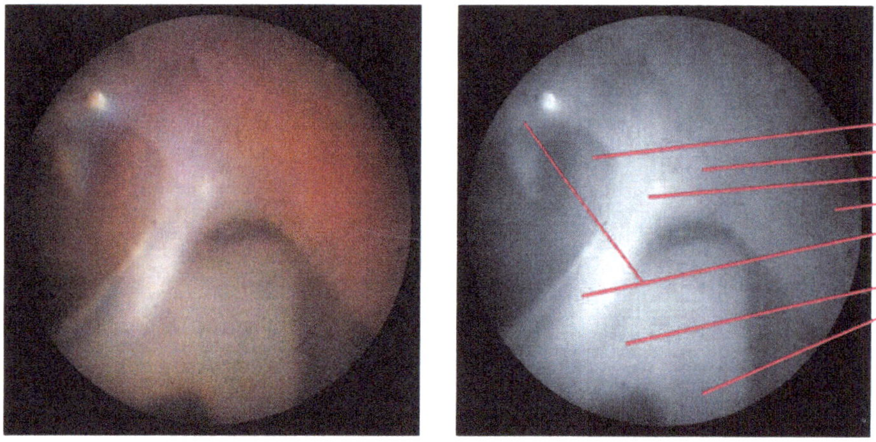

Fig. 197 View into the anterior pouch of TRÖLTSCH.

* Medial border of pouch of TRÖLTSCH

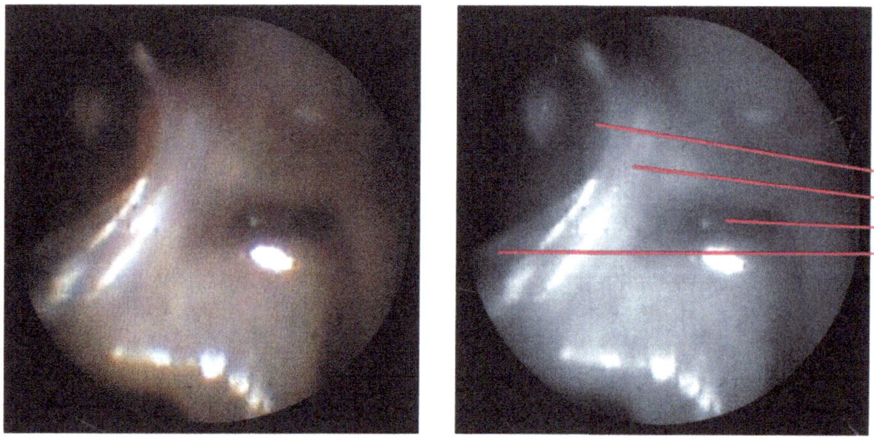

Fig. 198 Entrance to PRUSSAK's space.

D 1.3.3 Scope in mesotympanum bent laterally and downward: Hypotympanum

Fig. 199 The anterior inferior quadrant of the tympanic membrane is seen (left ear).

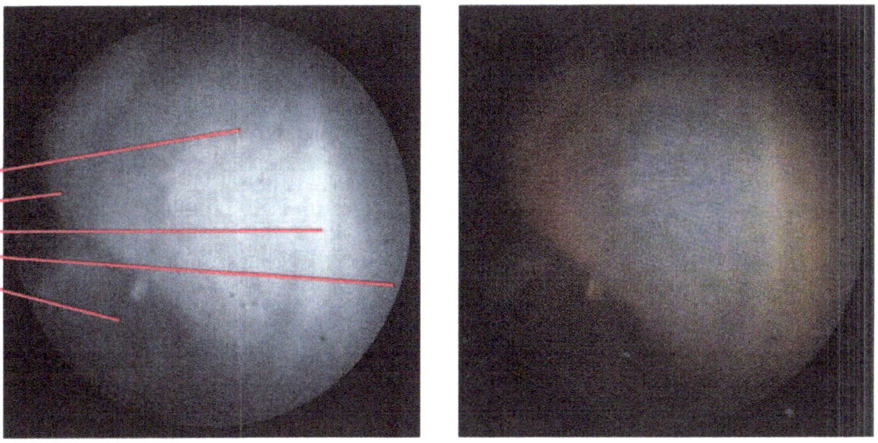

Fig. 200 Bending the flexible scope laterally and downward brings into view the hypotympanum.

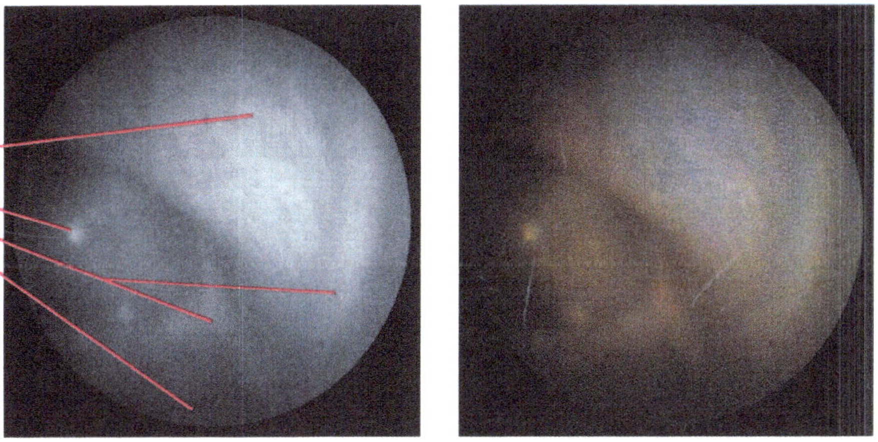

Fig. 201 Detail from Fig. 200 seen by moving the scope closer.

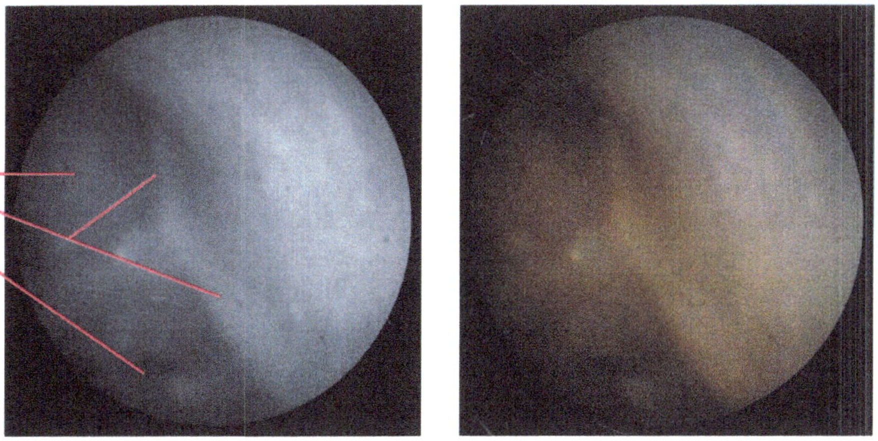

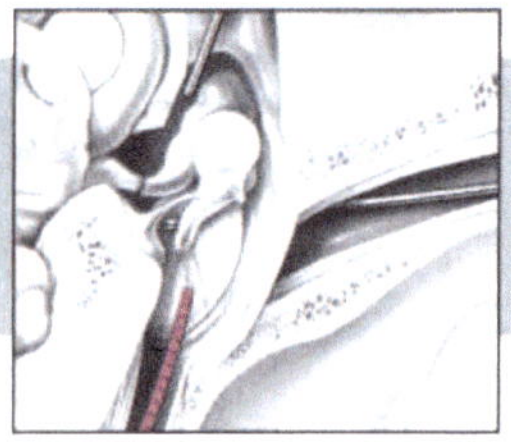

Inspection of epitympanum: Crossing over the tendon of the tensor tympani

D. 1.4.1 Scope in protympanum bent upward: Mucosal fold of tensor tympani muscle

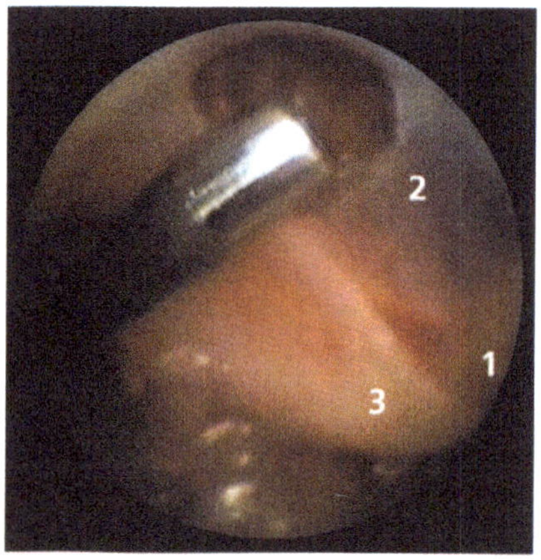

Fig. 202 Bent upward, the scope crosses over the tendon of the tensor tympani (1), pierces the mucosal fold (2) of the tensor tympani and enters the epitympanum. Cochleariform process (3). The tip of the scope lies at the level where it pierced the mucosal fold and is seen through a rigid 1.9 mm 30° scope in the auditory canal (left ear).

Images in Figs. 203-207 were obtained by gradually moving the transtubal scope forward into the epitympanum.

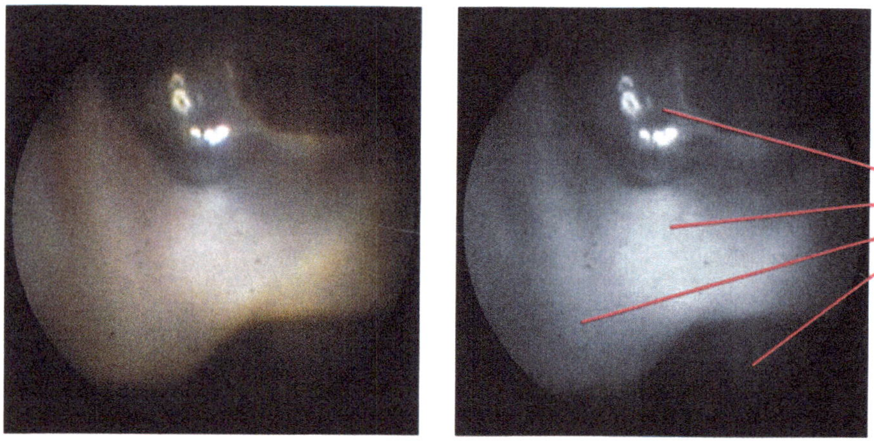

Fig. 203 Bending the scope upward brings into view the tendon of the tensor tympani.

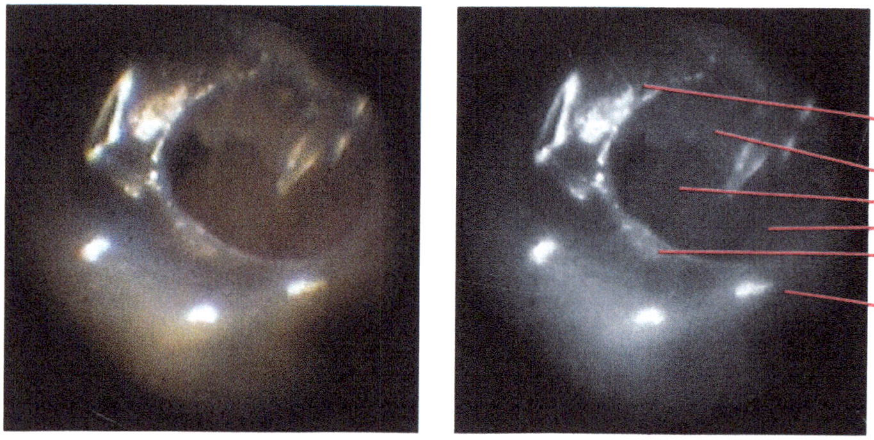

Fig. 204 Further bending offers a view through the mucosal fold of the tensor tympani into the epitympanum.

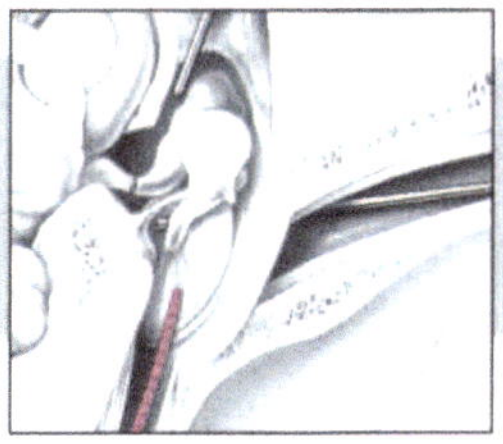

D 1.4.2 Scope advanced: Perforation of mucosal fold of tensor tympani

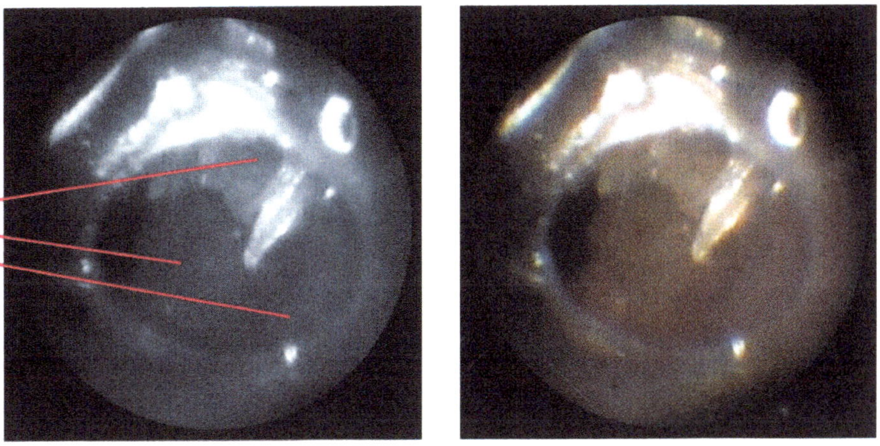

Fig. 205 Scope gradually advanced into the entrance to the epitympanum (left ear).

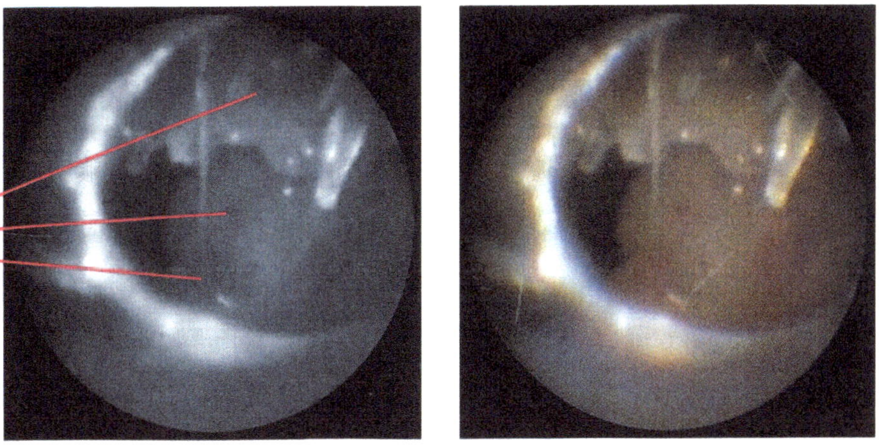

Fig. 206 Structures bordering the entrance to the epitympanum medially.

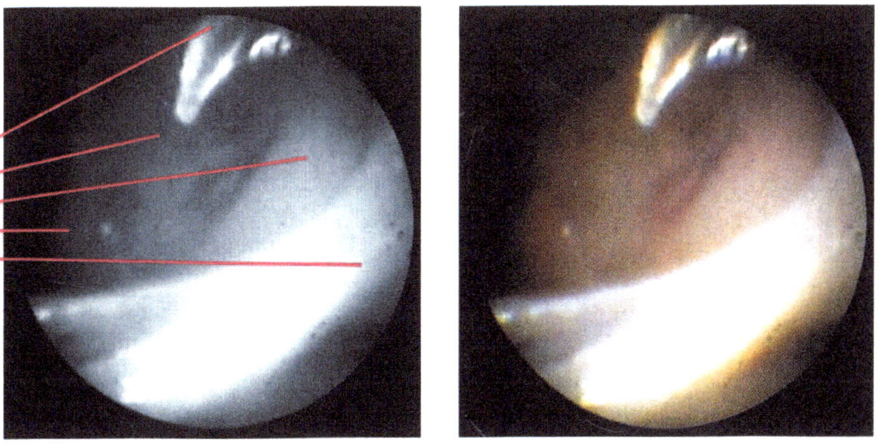

Fig. 207 Structures bordering the entrance to the epitympanum laterally.

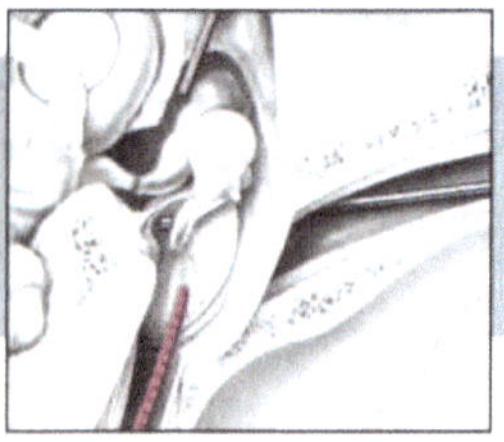

D 1.4.3 Scope above tendon of tensor tympani: Incudomallear joint

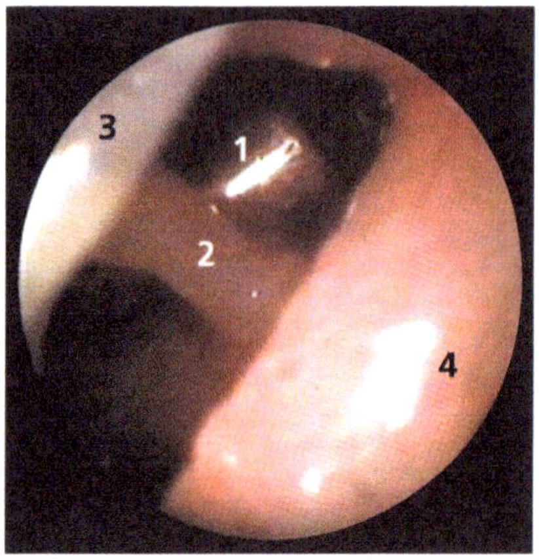

Fig. 208 Flexible scope (1) has crossed over the tendon of the tensor tympani (2) and entered the epitympanum. Through a rigid 1.9 mm 30° scope in the auditory canal its tip is seen to lie between the handle of the malleus (3) and the long limb of the incus (4) (left ear).

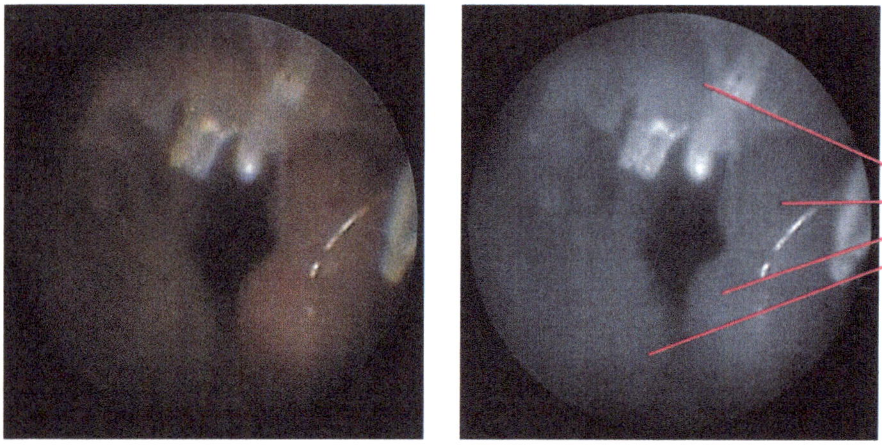

Fig. 209 View of the anterior epitympanum.

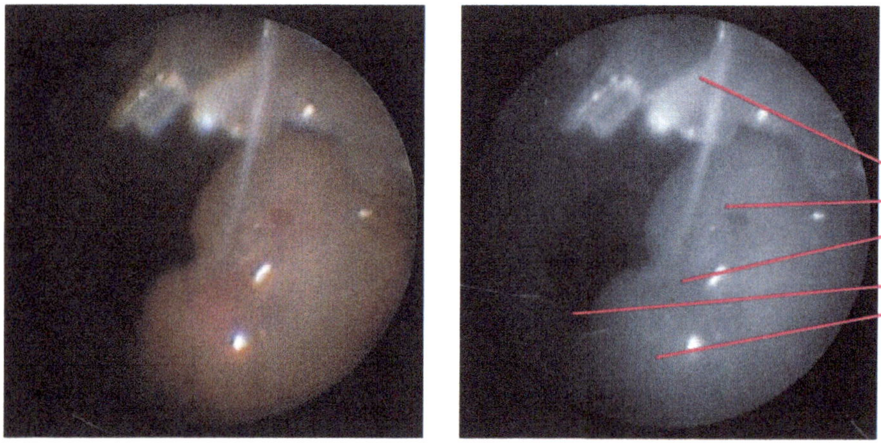

Fig. 210 The incudomallear joint is seen in the center.

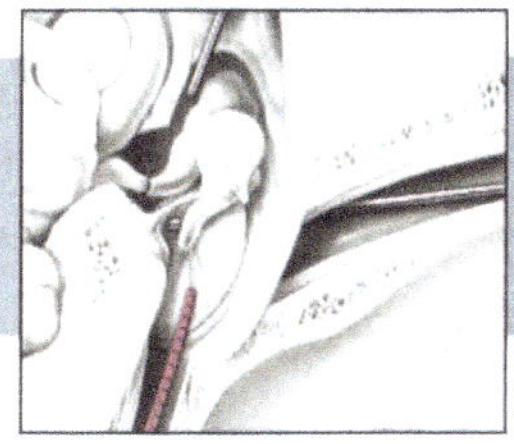

D 1.4.4 Scope medial to incudomallear joint: Posterior epitympanic recess

Fig. 211 View from the epitympanum down toward the mesotympanum (left ear).

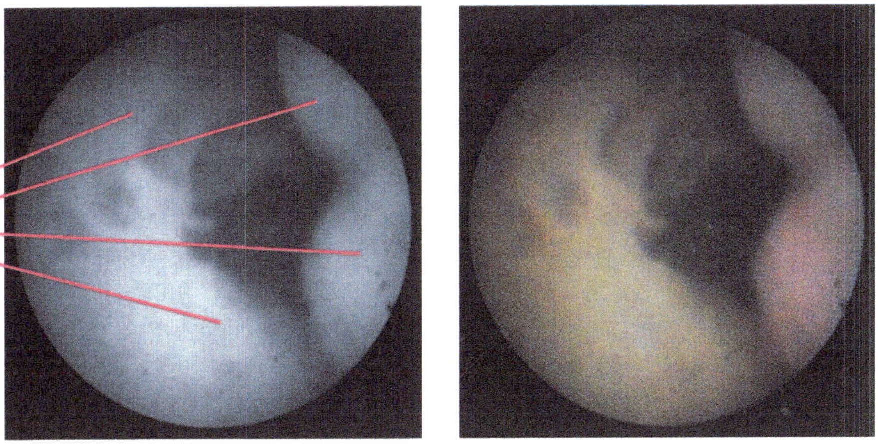

Fig. 212 Scope advanced toward the antrum mastoideum.

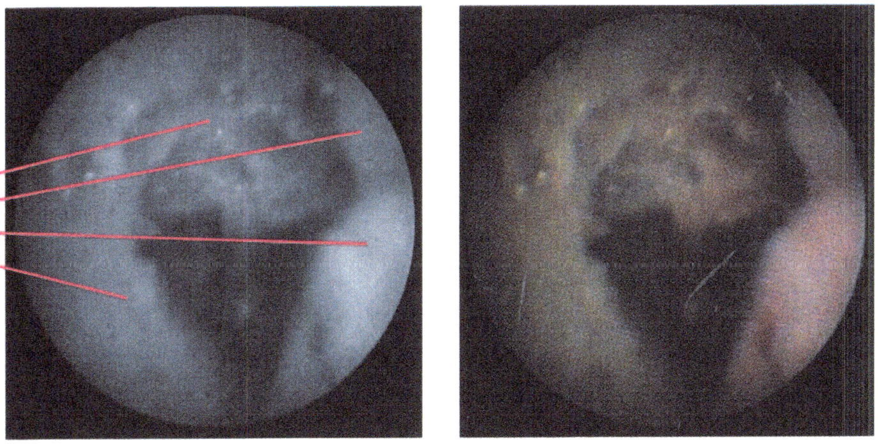

Fig. 213 Scope moved past the ossicular chain.

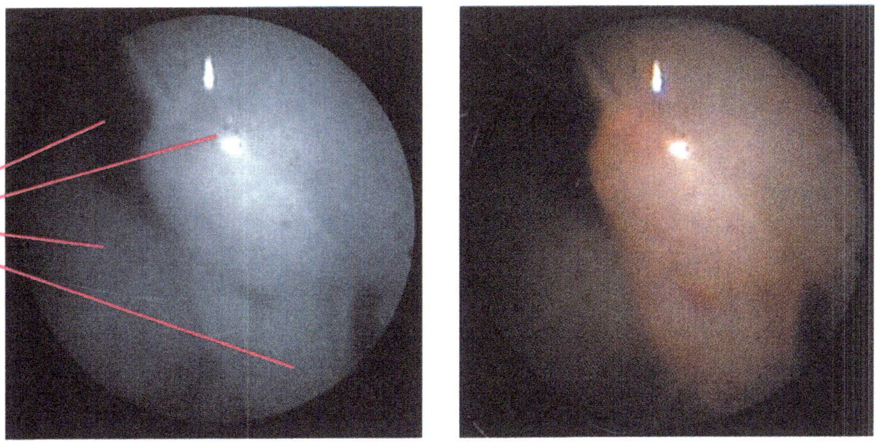

Inspection of antrum: Scope deep to the tendon of the tensor tympani, medial to incus

D 1.5.1 Scope below tendon of tensor tympani: Incudostapedial joint

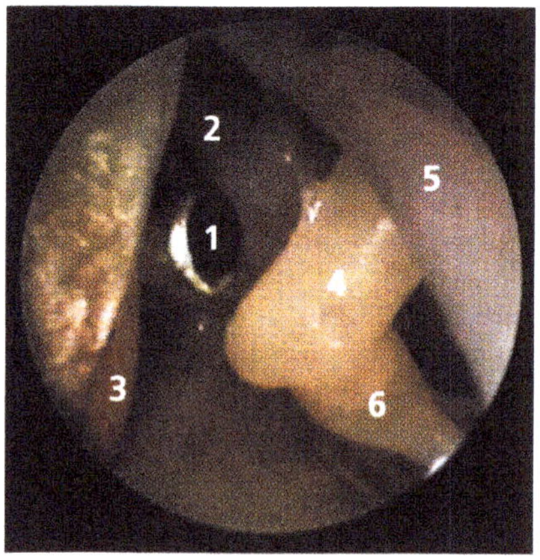

Fig. 214 The transtubal scope (1) is advanced toward the aditus ad antrum underneath the tendon of the tensor tympani (2). It is seen by the transmeatal scope between the handle of the malleus (3) and the long limb of the incus (4). The chorda tympani (5) and the head of the stapes (6) are also seen (left ear).

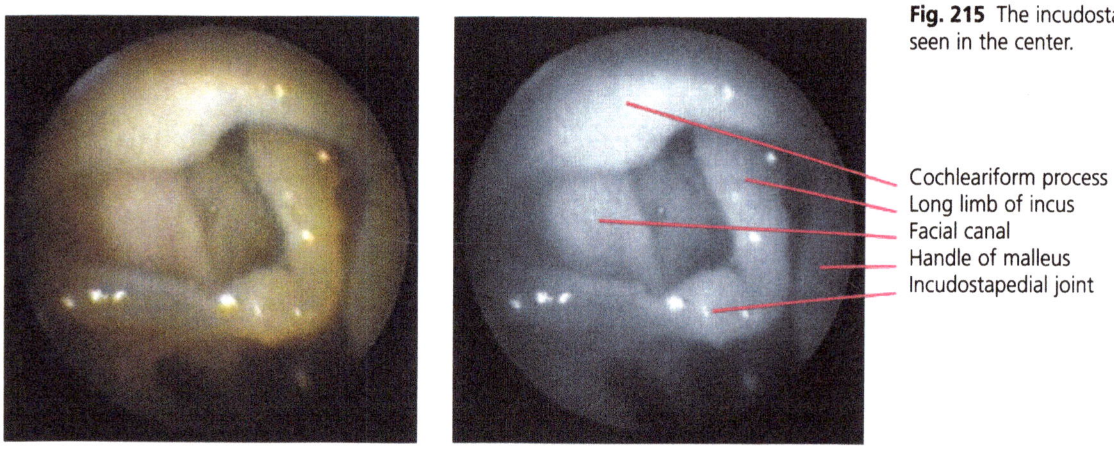

Fig. 215 The incudostapedial joint is seen in the center.

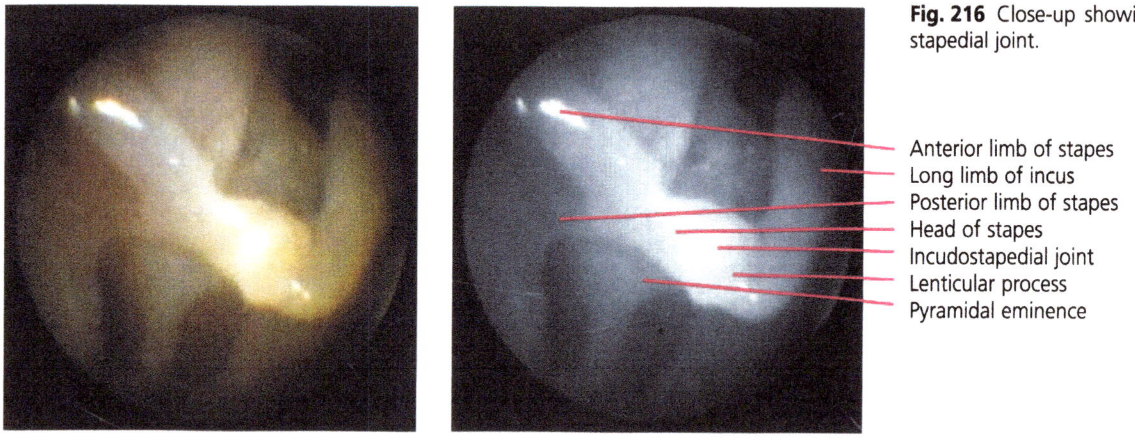

Fig. 216 Close-up showing the incudo-stapedial joint.

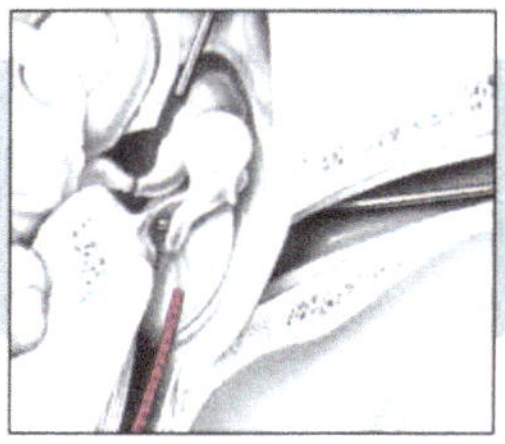

D 1.5.2 Scope medial to long limb of incus: Facial canal and lateral semicircular canal

Fig. 217 The flexible transtubal scope (1) crosses over the stapes (4) and is advanced into the gap between the long limb of the incus (2) and the facial canal (3) (left ear).

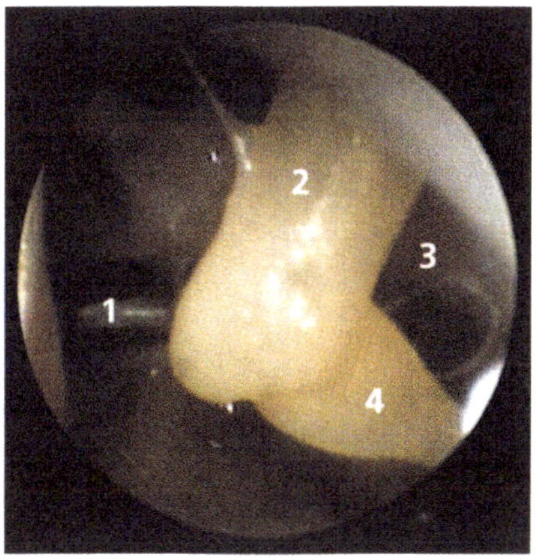

Fig. 218 View into the gap between the long limb of the incus and the facial canal with the scope looking toward the aditus ad antrum.

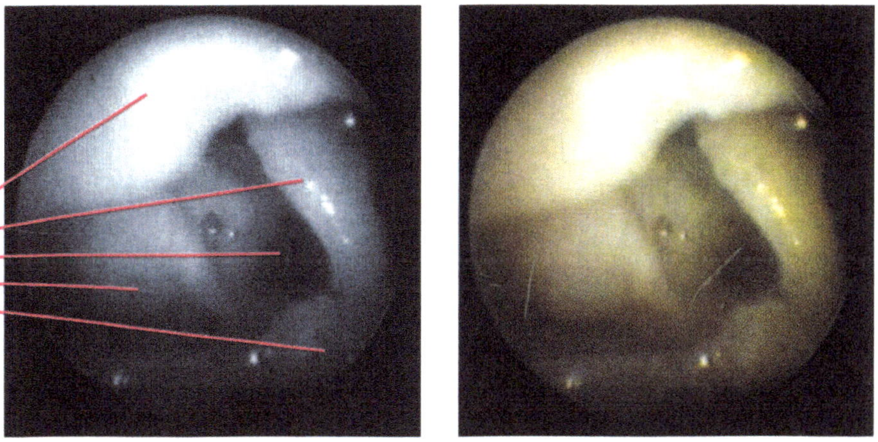

Fig. 219 Note the lateral semicircular canal in the center of the field of view.

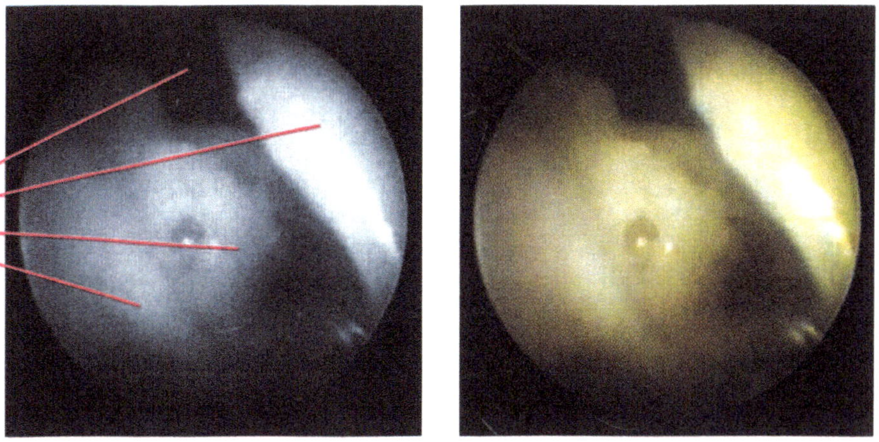

D 1.5.3 Scope crosses over stapes: Aditus ad antrum

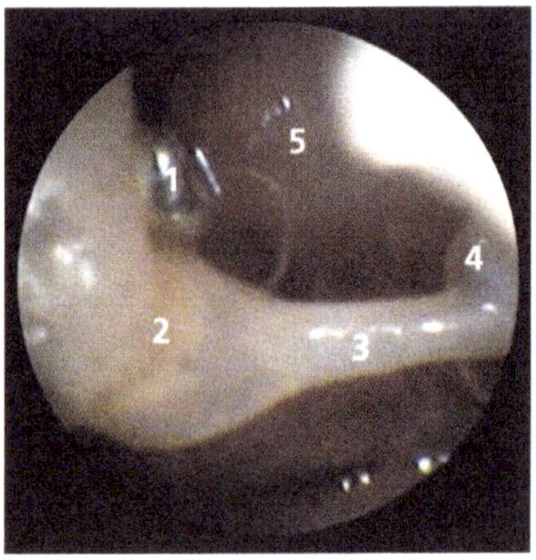

Fig. 220 The tip of the flexible scope (1) has crossed over the incudo-stapedial joint (2) and looks toward the antrum mastoideum. Tendon of stapedius (3), pyramidal eminence (4), facial canal (5) (left ear).

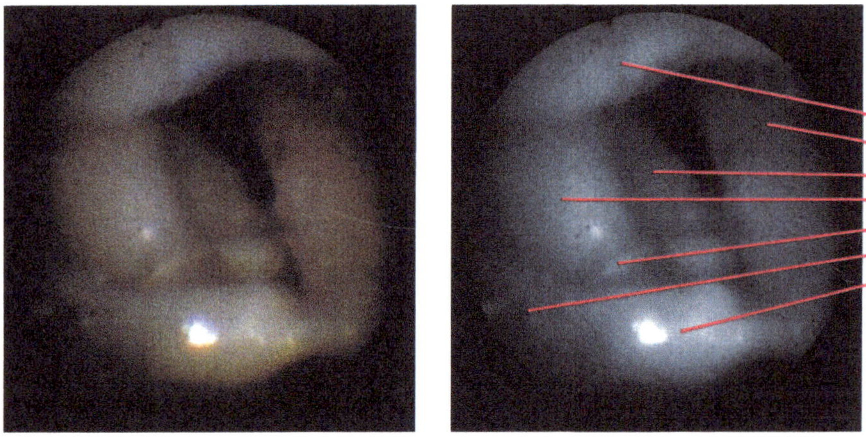

Fig. 221 View into the gap between the long limb of the incus and the facial canal.

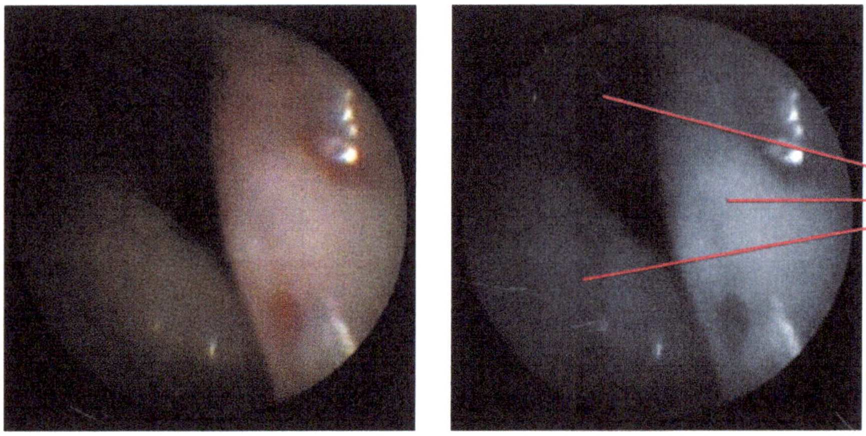

Fig. 222 Note narrow entrance to the aditus ad antrum between the long limb of the incus and the lateral semicircular canal (compare Fig. 189).

Inspection of chorda tympani: Scope deep to the tendon of the tensor tympani lateral to incus

D 1.6.1 Scope bent laterally and upward behind handle of malleus: Chorda tympani

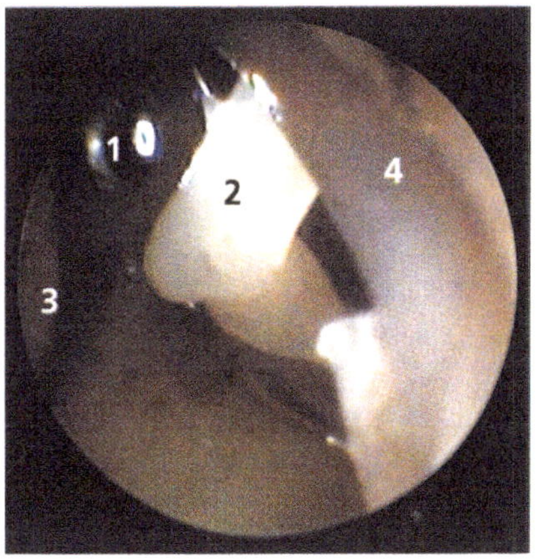

Fig. 223 The transtubal scope (1) is bent laterally and upward and comes to lie lateral to the long limb of the incus (2) and medial to the handle of the malleus (3). The chorda tympani (4) is seen in the periphery (left ear).

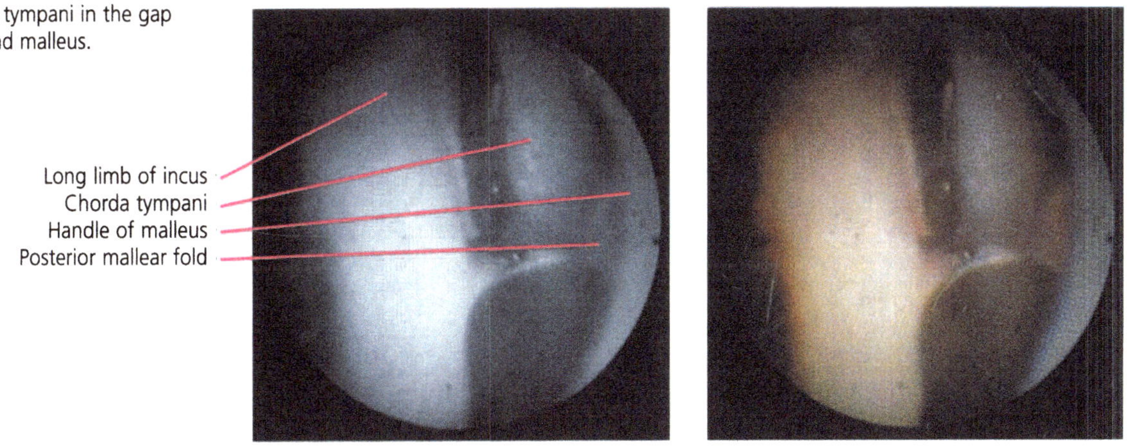

Fig. 224 Chorda tympani in the gap between incus and malleus.

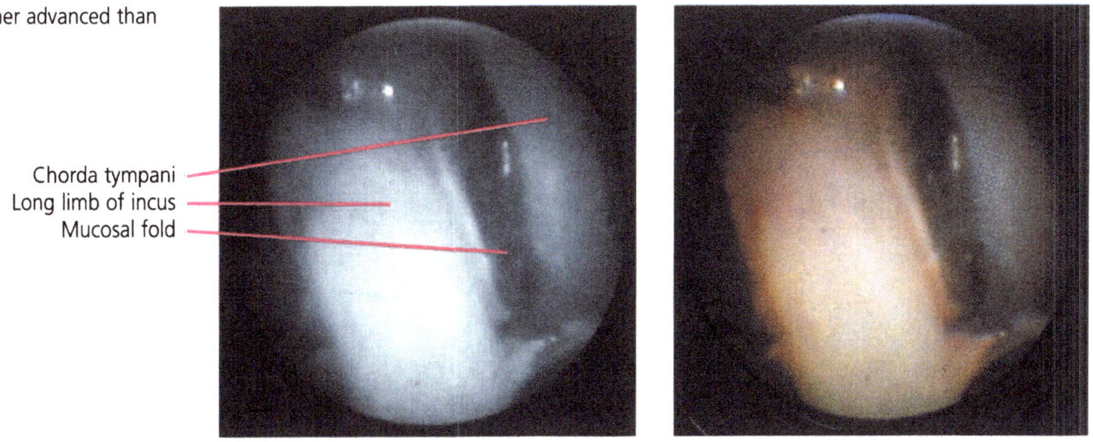

Fig. 225 Scope further advanced than in Fig. 224.

D 1.6.2 Scope lateral to long limb of incus: Course of chorda tympani

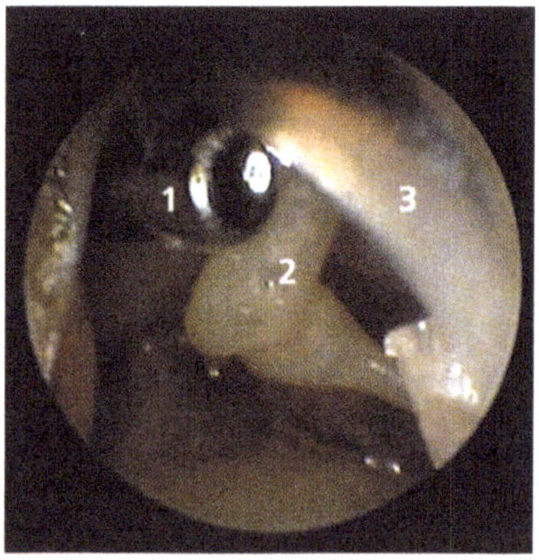

Fig. 226 The transtubal scope (1) is moved ahead laterally to the long limb of incus (2) and looks at the chorda tympani (3) (left ear).

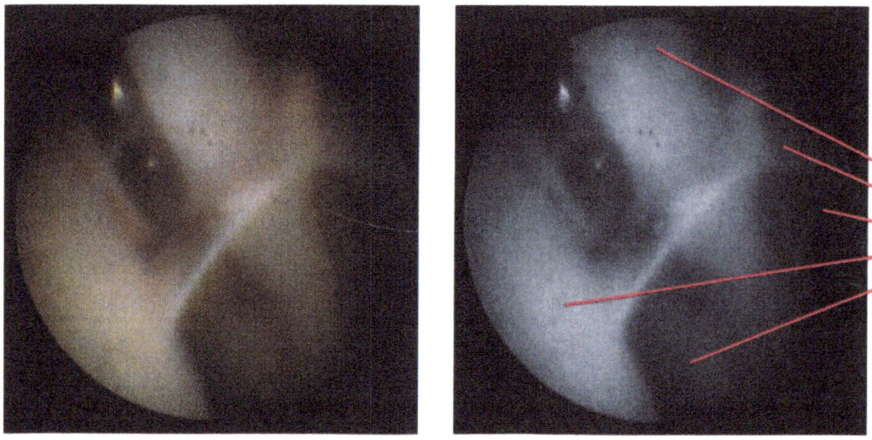

Fig. 227 View of the posterior inferior segment of the chorda tympani.

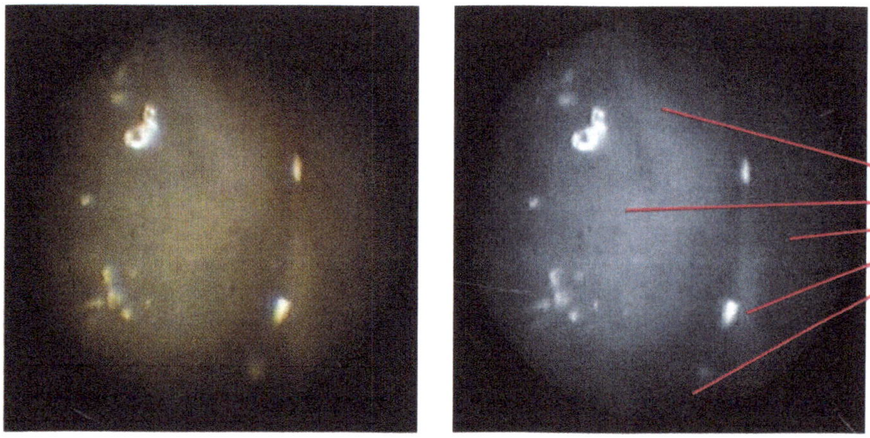

Fig. 228 View of the chorda tympani at its emergence from the eminence.

Inspection of posterior wall: Scope deep to the tendon of the tensor tympani and stapes

D 1.7.1 Scope in the plane of the handle of malleus: stapes

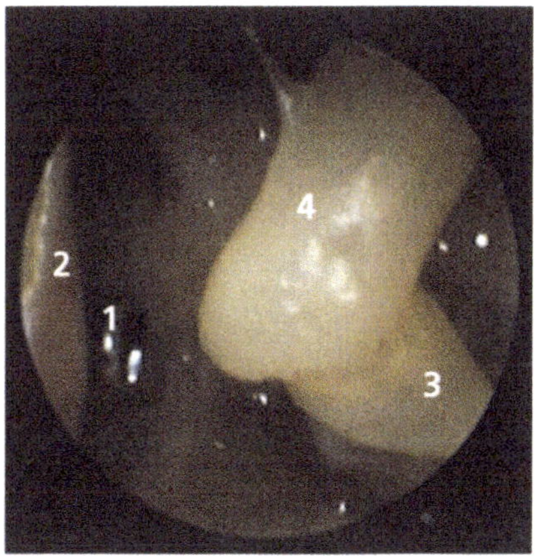

Fig. 229 The transtubal scope (1) is medial to the handle of malleus (2) and slightly bent to look onto the stapes (3); incus (4); (left ear).

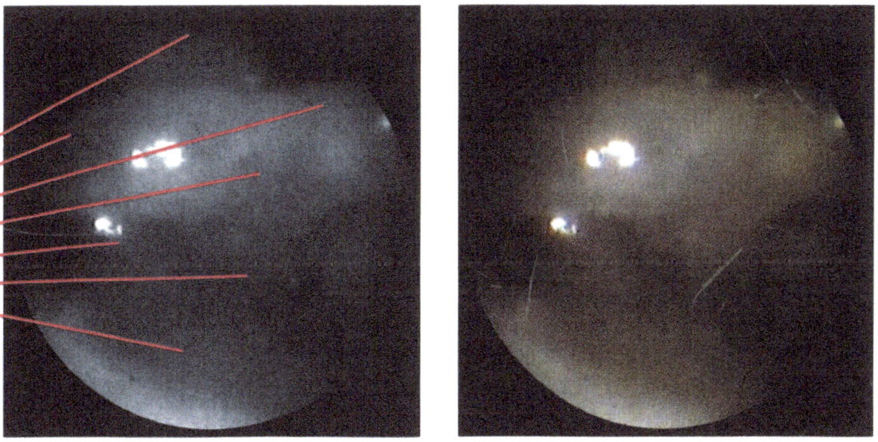

Fig. 230 Scope focuses on the stapes.

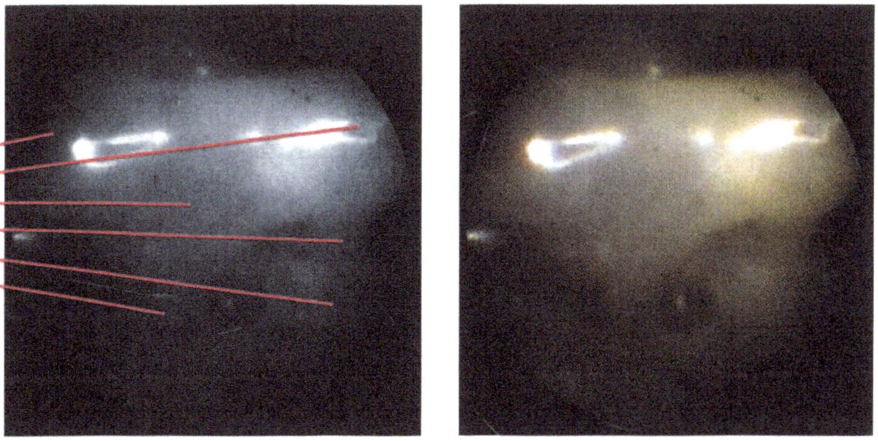

Fig. 231 Stapes and pyramidal eminence.

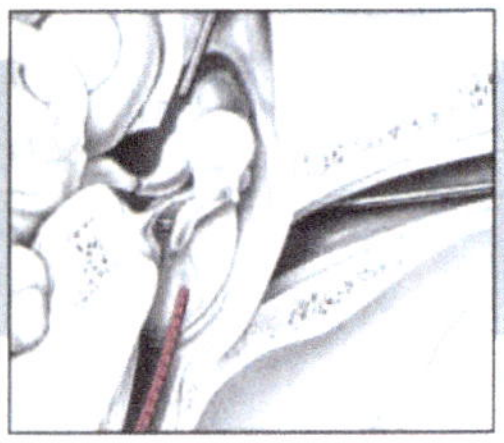

D 1.7.2 Scope below incudostapedial joint: Posterior wall

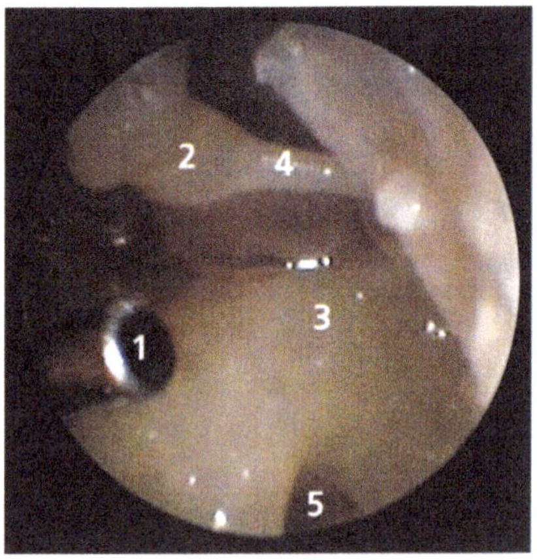

Fig. 232 The transtubal scope (1) is below the incudostapedial joint (2) and lateral to the promontory (3) and looks toward the tendon of the stapedius (4) and the pyramidal eminence. It is seen through a rigid 1.9 mm 30° scope in the auditory canal. Round window niche (5); (left ear).

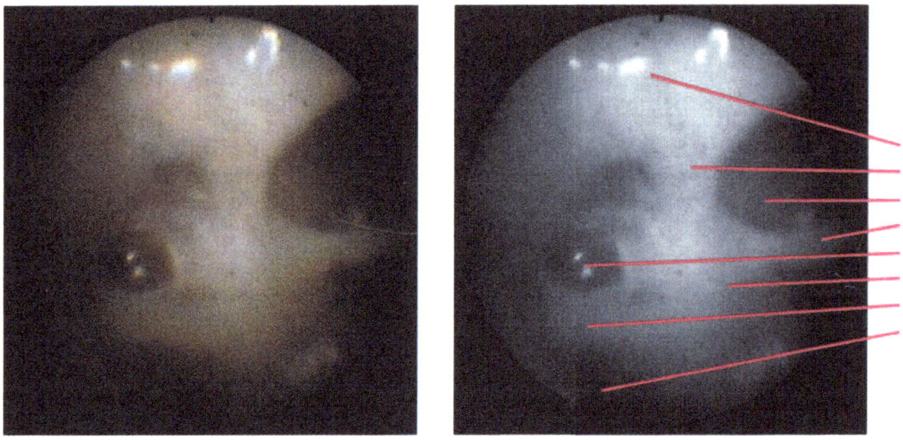

Fig. 233 Pyramidal eminence and adjacent structures are seen in the center of the field of view.

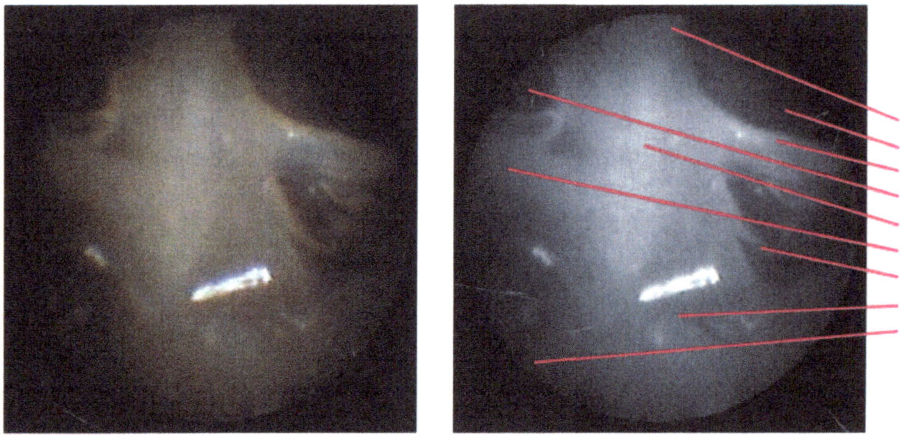

Fig. 234 Scope looks slightly more lateral than in Fig. 233.

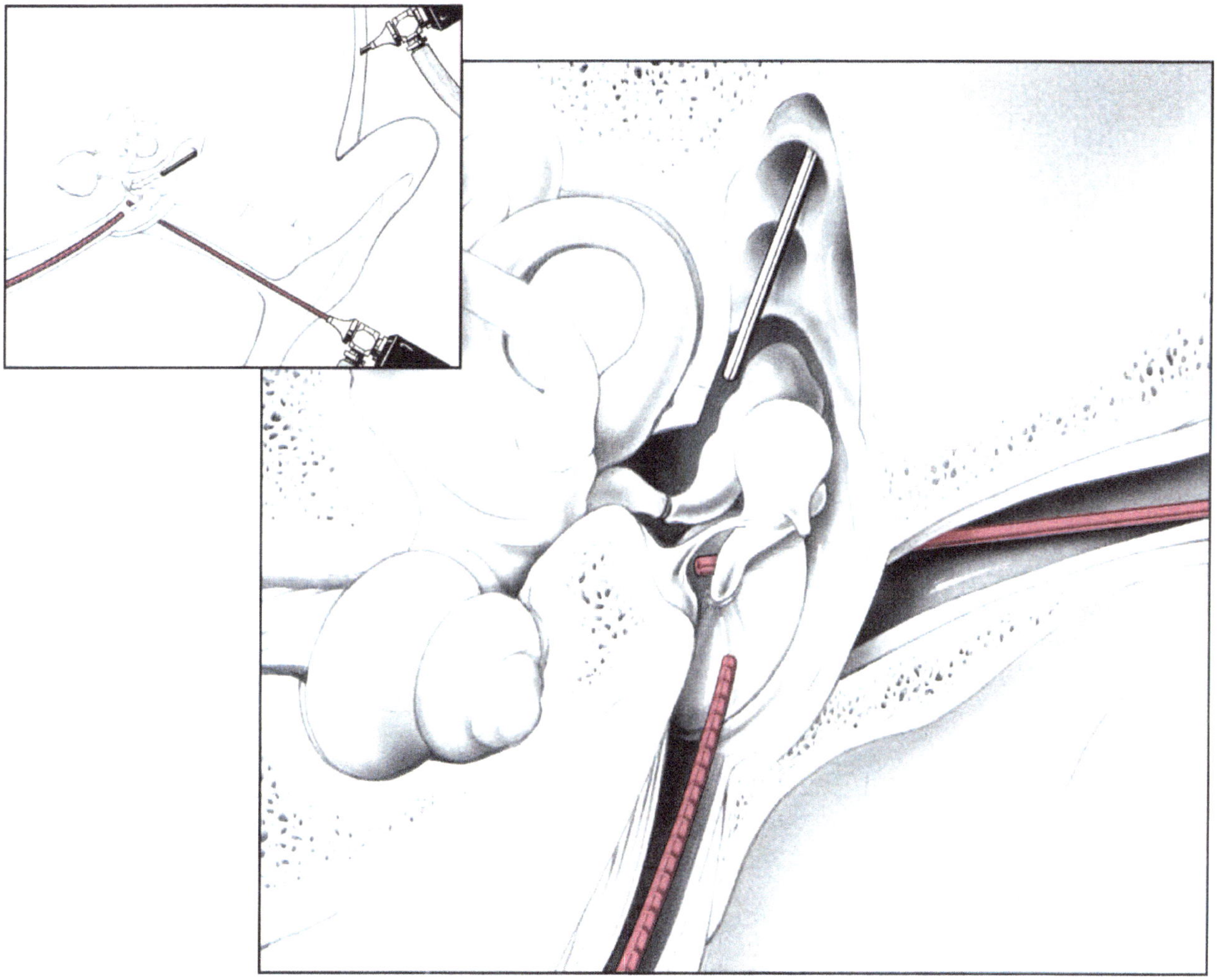

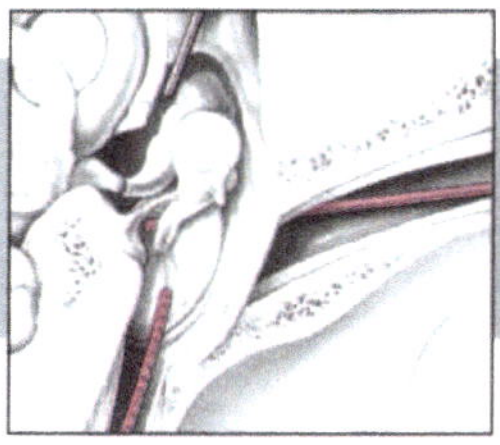

Two-port approach to the middle ear

Along the transtympanic and the transmeatal approach the lateral wall of the middle ear cannot be seen because of the "keyhole effect". The incudomallear joint and the chorda tympani block the view of the epitympanum. But both the lateral wall and the epitympanum can be explored through a transtubal scope. In the two-port approach the 2 scopes are arranged at an angle of about 70° to one another. As a result (1) the dead angle of one scope can be seen by the other scope and (2) maneuvers of one scope - particularly the flexible steerable scope - can be monitored by the other scope. With two-port endoscopy of the middle ear the entire tympanic cavity can be explored. Segments of the middle ear, which cannot be seen by one scope, are accessible to inspection by the other.

Why different scopes (30° and 70°) are used for approaching the middle ear through the external auditory canal is discussed at length in the sections on the transmeatal approach (Section A) and the transtympanic approach (Section B). Section D gives details about the transtubal approach.

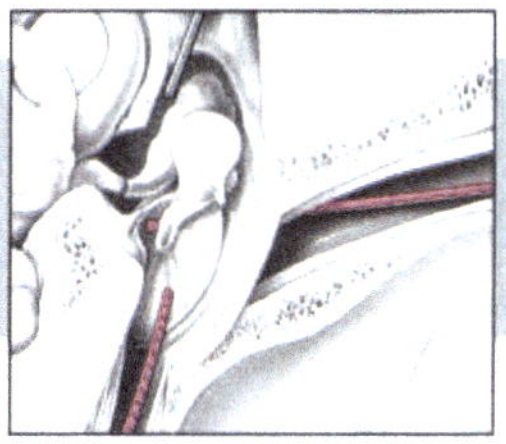

Two-port approach

1.9 mm 30° rigid scope —
0.8 mm flexible steerable scope

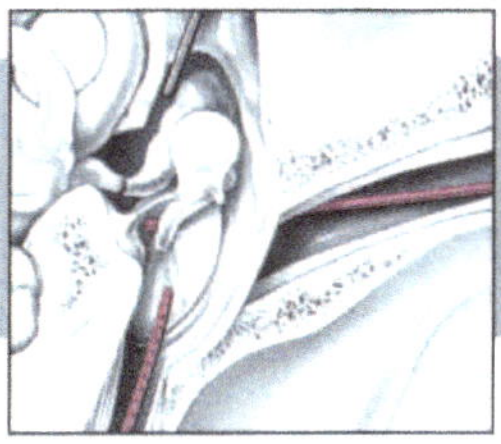

E 1.1 Transtubal scope crosses above the tendon of the tensor tympani

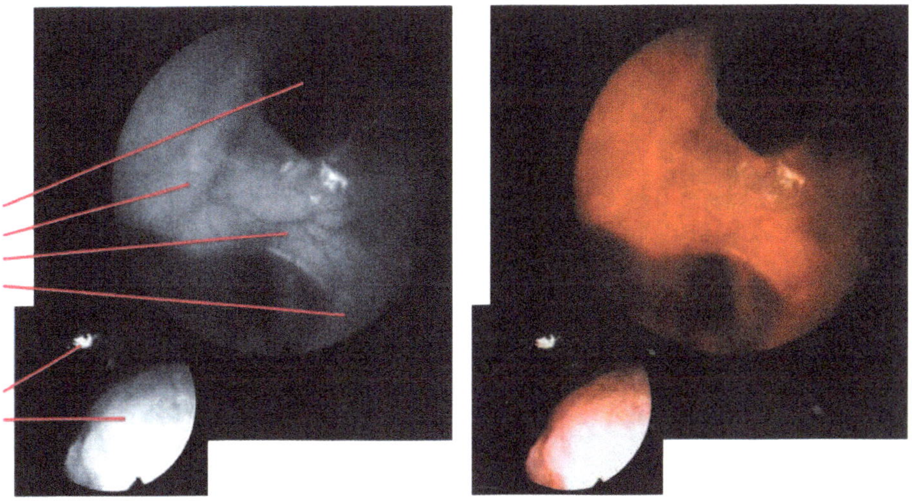

Fig. 235 The flexible and steerable transtubal scope lies in the protympanum and looks at the tensor tympani muscle, as it emerges from the cochleariform process. Through the rigid scope the relation of the flexible scope to the promontory is seen (rigid scope, 1.9 mm, 30°, left ear).

Mucosal fold
Cochleariform process
Tendon of tensor tympani
Handle of malleus
Fiberscope
Promontory

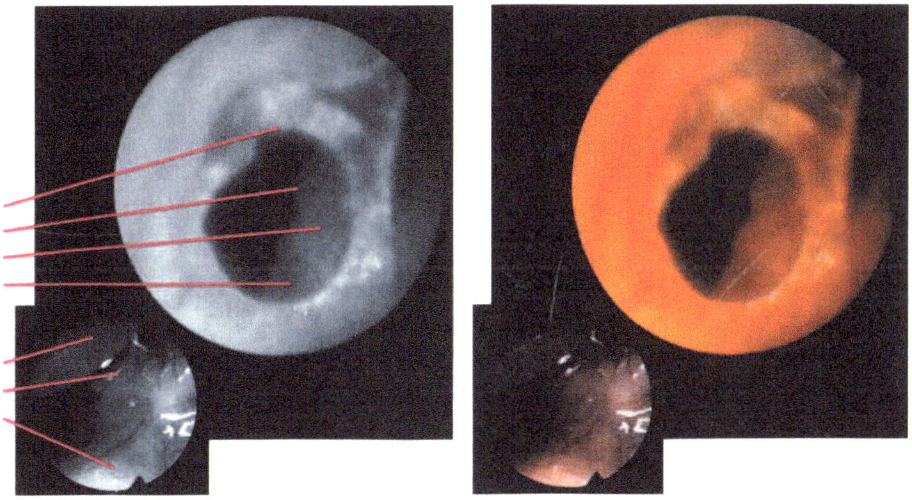

Fig. 236 The fiberscope is bent upwards to look into the epitympanic recess through a perforation in the mucosal fold of the tensor tympani muscle. The monitoring scope (rigid, 1.9 mm, 30°) shows the tip of the fiberscope just above the tendon of the tensor tympani.

Perforation in the mucosal fold
Head of malleus
Incudomallear joint
Body of incus
Handle of malleus
Cochleariform process
Promontory

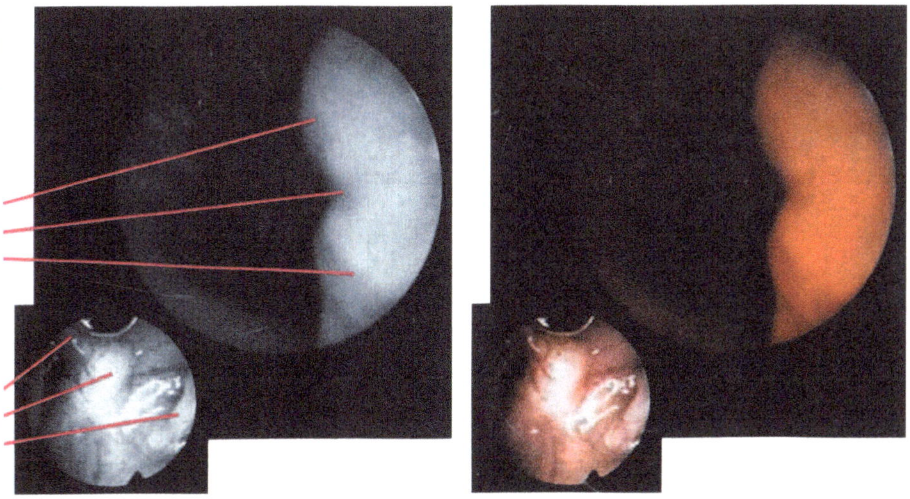

Fig. 237 The scope is advanced into the anterior epitympanic recess offering a view of the incudomallear joint. The monitoring scope sees the tip of the fiberscope in the anterior tympanic isthmus (rigid scope, 1.9 mm, 30°).

Head of malleus
Incudomallear joint
Body of incus
Tendon of tensor tympani
Cochleariform process
Anterior limb of stapes

E 1.2 Transtubal scope crosses below the tendon of the tensor tympani

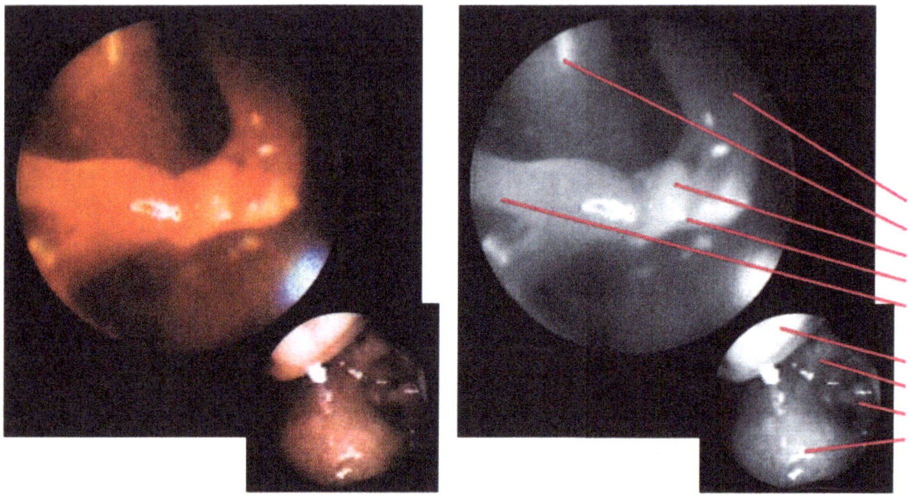

Fig. 238 Unbent the transtubal scope can be advanced below the tendon of the tensor tympani to look at the incudo-stapedial joint. The transmeatal scope shows its relation to the cochleariform process and the remaining distance to the stapes (rigid scope, 1.9 mm, 30°, left ear).

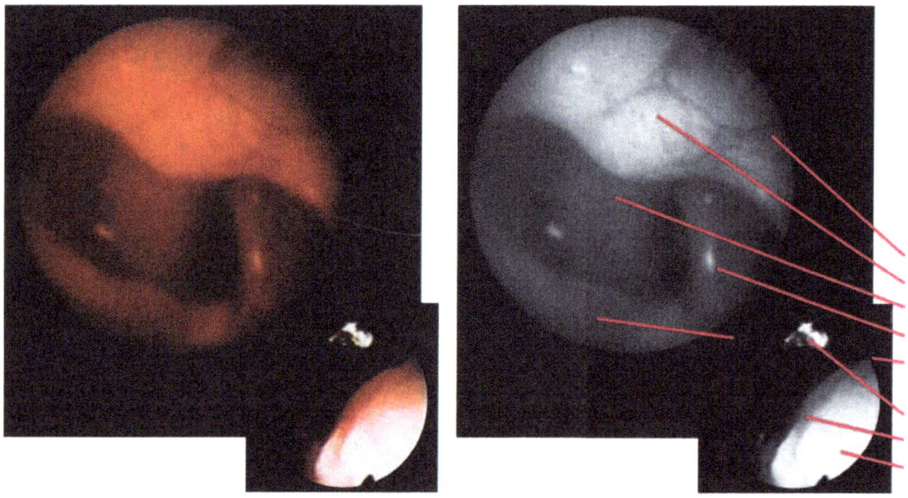

Fig. 239 The fiberscope is bent upward to look toward the aditus ad antrum. This also brings into view the cochleariform process in the upper part of the image. Through the transmeatal scope (rigid, 1.9 mm, 30°) the tip of the fiberscope is seen to lie in the tympanic cavity close to JACOBSON's nerve.

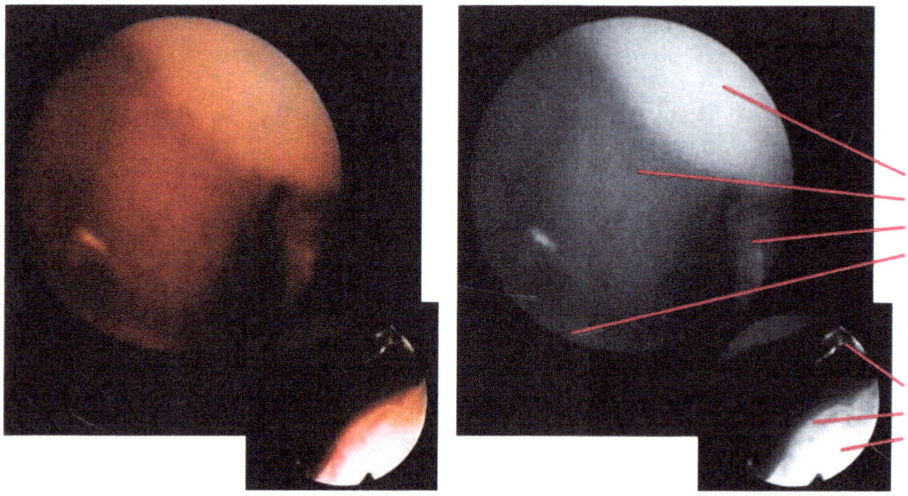

Fig. 240 With the transtubal fiberscope moved on, a very prominent facial canal is seen which obstructs further progress toward the aditus ad antrum (rigid scope, 1.9 mm, 30°).

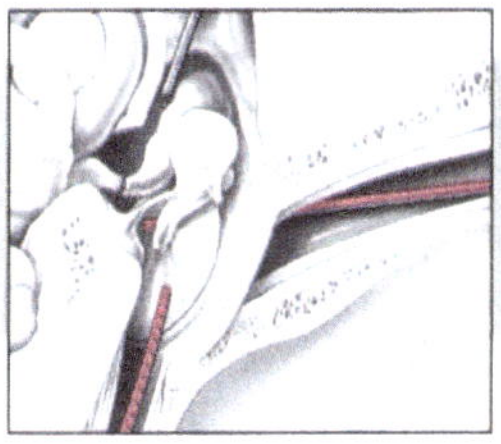

E 1.3 Transtubal scope looks at the ossicular chain

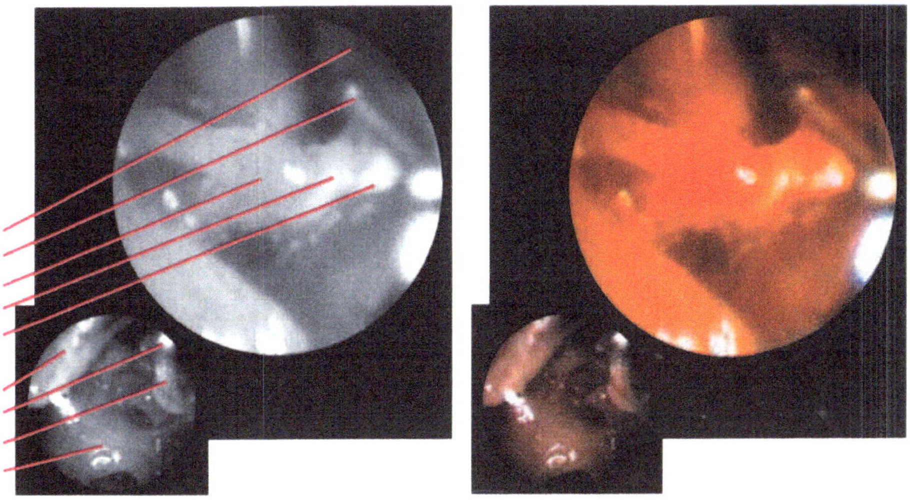

Fig. 241 Lying below the tendon of the tensor tympani the transtubal fiberscope looks at the incudostapedial joint. Manipulations of the joint by a little hook are well seen in an anteroposterior view. The transmeatal scope (rigid, 1.9 mm, 30°) was used to monitor the insertion of the hook (left ear).

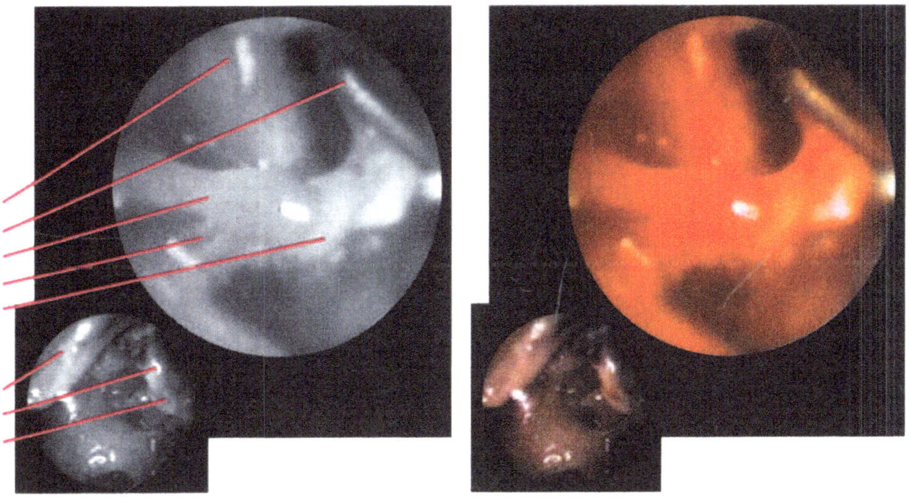

Fig. 242 Transtubal scope slightly closer to the incudostapedial joint: Pulling with the hook shows the incudostapedial joint to move freely. The stapes suprastructure is also seen to move (rigid scope, 1.9 mm, 30°).

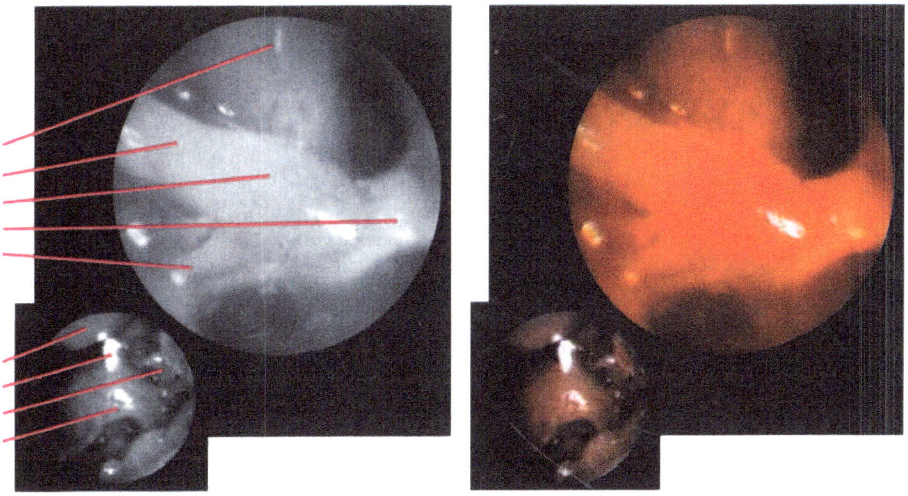

Fig. 243 Close-up view of stapes through the transtubal fiberscope monitored through a transtympanic scope (rigid, 1.9 mm, 30°).

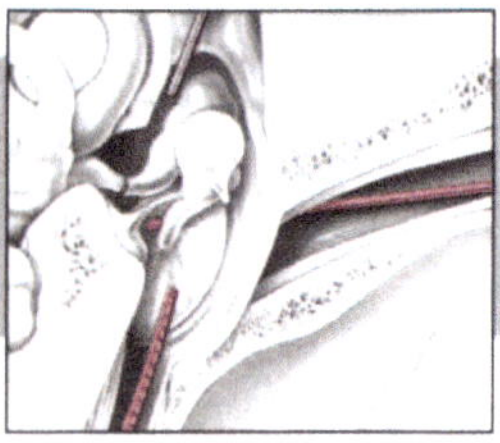

E 1.4 Transtubal scope looks at the posterior wall

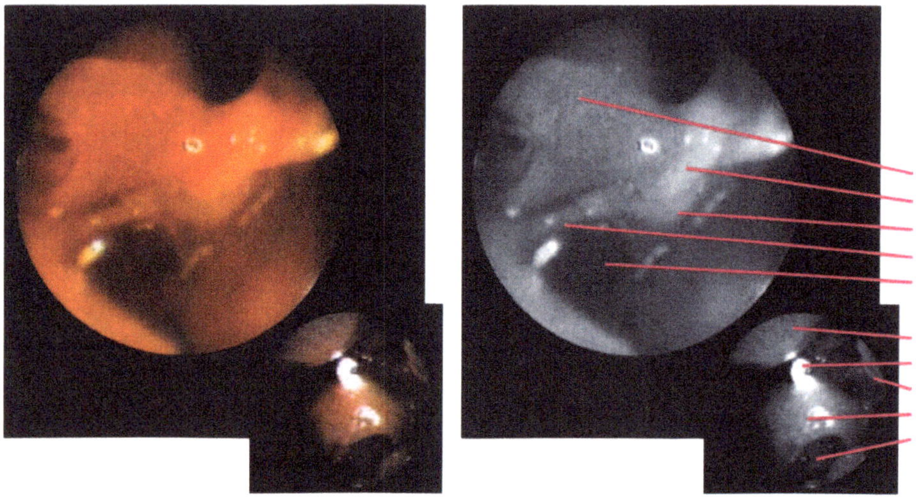

Fig. 244 In this sequence the transtubal fiberscope is bent downward to explore the posterior wall. The transmeatal scope shows its relation to the tympanic structures (rigid scope, 1.9 mm, 30°, left ear).

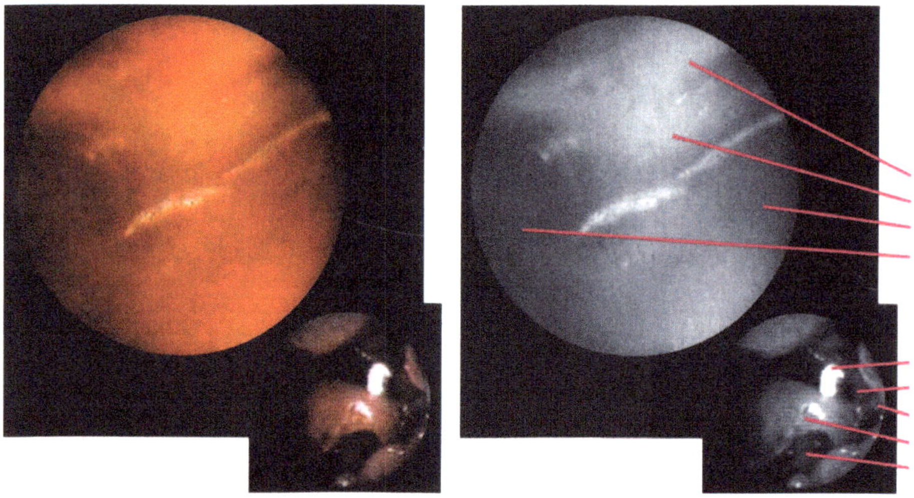

Fig. 245 With the fiberscope advanced a close-up view of the pyramidal eminence is obtained. The transmeatal scope shows the close relation of the fiberscope to the stapes (rigid scope, 1.9 mm, 30°).

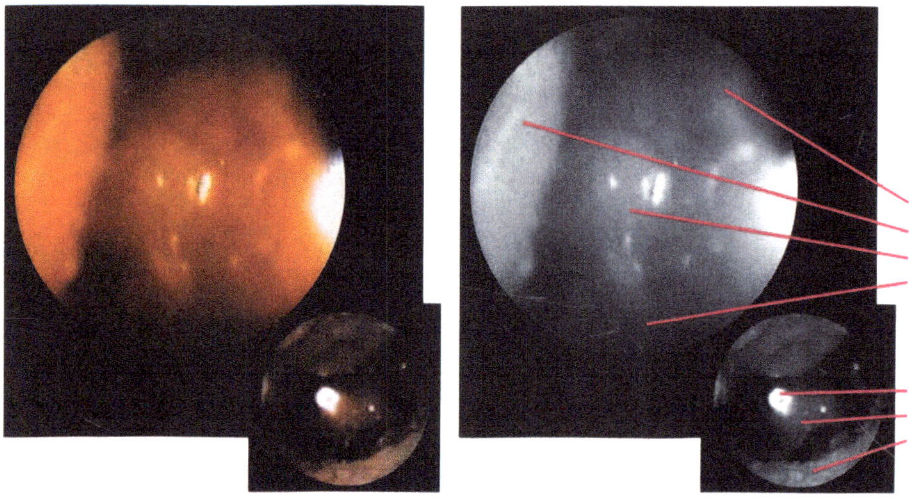

Fig. 246 Further bending of the fiberscope provides a view of the inferior part of the mastoid wall. Through the transmeatal scope (rigid, 1.9 mm, 30°) the tip of the fiberscope is seen near the mid-promontory.

Index

SpringerMedicine

Peter S. Hechl, Reuben C. Setliff III, Manfred Tschabitscher

Endoscopic Anatomy of the Paranasal Sinuses

1997. XV, 130 pages. 194 partly coloured figures.
Hardcover DM 298,–, öS 2086,–
(recommended retail price)
ISBN 3-211-82922-9

For the beginner or for the accomplished sinus surgeon, mastering the anatomy of the lateral nasal wall is an ongoing challenge. Even though there are excellent standard anatomical references and equally outstanding sinus courses with cadaver dissection, a reference depicting the surgical anatomy is needed. A step-by-step surgical approach on the anterior nasal spine to the anterior wall of the sphenoid is presented. The sinus surgeon is confronted with a wide range of different spaces created by the ethmoid bone. No other bone in the human body has so many anatomical variations. Four critical anatomical structures are emphasized as the foundation for a precise approach to surgery of the maxillary, anterior ethmoid, frontal, and posterior ethmoid sinuses. The goal of this book is to meet the tremendous challenge of offering an anatomical approach which will serve the sinus surgeon of every level of experience and expertise.

"... an atlas of surgical endonasal anatomy comprising exhaustive illustrations of very high quality. The legends are clearly appended ... The colors and contrasts which are important aids in endoscopy are perfectly presented ... this atlas appears very useful for understanding the anatomy of the ethmoidal labyrinth. It will familiarise the surgeon's eye with the fundamental anatomic landmarks in normal and pathologic conditions ..."

Surgical and Radiologic Anatomy

"... an important aid to the acquisition of endoscopic knowledge and ability ..."

Australian Journal of Otolaryngology

"... a leading anatomical reference book for planning of endoscopic procedures involving the paranasal sinuses and therefore this book should prove useful to ENT-surgeons as well as to neurosurgeons ..."

Minimally Invasive Neurosurgery

SpringerWienNewYork

A-1201 Wien, Sachsenplatz 4–6, P.O.Box 89, Fax +43.1.330 24 26, e-mail: books@springer.at, Internet: **www.springer.at**
D-69126 Heidelberg, Haberstraße 7, Fax +49.6221.345-229, e-mail: orders@springer.de
USA, Secaucus, NJ 07096-2485, P.O. Box 2485, Fax +1.201.348-4505, e-mail: orders@springer-ny.com
Eastern Book Service, Japan, Tokyo 113, 3–13, Hongo 3-chome, Bunkyo-ku, Fax +81.3.38 18 08 64, e-mail: orders@svt-ebs.co.jp

GPSR Compliance

The European Union's (EU) General Product Safety Regulation (GPSR) is a set of rules that requires consumer products to be safe and our obligations to ensure this.

If you have any concerns about our products, you can contact us on ProductSafety@springernature.com

In case Publisher is established outside the EU, the EU authorized representative is:

Springer Nature Customer Service Center GmbH
Europaplatz 3
69115 Heidelberg, Germany

Batch number: 10731490

Printed by Printforce, the Netherlands